SYSTEMS ENGINEERING HANDBOOK

A GUIDE FOR SYSTEM LIFE CYCLE PROCESSES AND ACTIVITIES

INCOSE-TP-2003-002-03.2.1
January 2011

Prepared by:
SE Handbook Working Group
International Council on Systems Engineering (INCOSE)
7670 Opportunity Rd, Suite 220
San Diego, CA 92111-2222

Edited by:
Cecilia Haskins, CSEP

Revisions and Appendices D to N of v3.1 edited by:
Kevin Forsberg, CSEP
Michael Krueger, CSEP

Revisions and Updates to v3.2 and 3.2.1 edited by:
Michael Krueger, CSEP
David Walden, CSEP
R. Douglas Hamelin

INCOSE Notices

This INCOSE Technical Product was prepared by the SE Handbook Working Group of the International Council on Systems Engineering (INCOSE). It is approved by INCOSE Technical Operations Leadership for release as an INCOSE Technical Product.

ISBN 0-9720562-9-7

History of Changes

Revision	Revision Date	Change Description & Rationale
Original	1994	Original by INCOSE members from several defense/aerospace companies, including Lockheed, TRW, Northrop Grumman, Ford Aerospace, CSM; Heavy US DoD bias.
1.0	5/19/1999	Revision approved to update and broaden coverage of SE process; broad participation of INCOSE members as authors
2.0	4/1/2002	Revision approved to expand on several topics such as Functional Analysis; included new copyright approval. This version was the basis for the first CSEP exam
2.0A	6/1/2004	Reduced page count of SEH v2 by 25% based on complaints from users and CSEP candidates about the uneven treatment of topics and excessively long sections on some areas (like the 70 pages on functional analysis). It still had a heavy US DoD bias.
3.0	3/20/2006	First version to be based on ISO/IEC 15288:2002; the intent was to move away from the US DoD emphasis. This was to be a small written document.
3.1	8/15/2007	Added detail which was not previously published. This is the version that was submitted to ISO/IEC for consideration as an ISO/IEC Technical Report
3.2	2/1/2010	Updated version based on ISO/IEC 15288:2008; resubmitted to ISO/IEC for consideration as an ISO/IEC Technical Report
3.2.1	1/30/2011	Clarified definitional material, architectural frameworks, concept of operations references, risk references and editorial corrections based on ISO/IEC review. To be available as ISO TR 16337 (available mid-2011)

Preface

The objective of the International Council on Systems Engineering (INCOSE) *Systems Engineering Handbook* is to provide a description of key process activities performed by systems engineers. The intended audience is the new systems engineer, an engineer in another discipline who needs to perform systems engineering (SE), or an experienced systems engineer who needs a convenient reference. The primary purpose of the version 3.2.1 update is to:

1. Resolve ISO/IEC balloting comments
2. Resolve inconsistencies and errata in v3.2.

The descriptions in this handbook show what each SE process activity entails, in the context of designing for affordability and performance. On some projects, a given activity may be performed very informally; on other projects, very formally, with interim products under formal configuration control. This document is not intended to advocate any level of formality as necessary or appropriate in all situations. The appropriate degree of formality in the execution of any SE process activity is determined by:

1. The need for communication of what is being done (across members of a project team, across organizations, or over time to support future activities)
2. The level of uncertainty
3. The degree of complexity
4. The consequences to human welfare.

On smaller projects, where the span of required communications is small (few people and short project life cycle) and the cost of rework is low, SE activities can be conducted very informally (and thus at low cost). On larger projects, where the cost of failure or rework is high, increased formality can significantly help in achieving project opportunities and in mitigating project risk.

In a project environment, work necessary to accomplish project objectives is considered “in scope;” all other work is considered “out of scope.” On every project, “thinking” is always “in scope.” Thoughtful tailoring and intelligent application of the SE process described in this handbook is essential to achieve the proper balance between the risk of missing project technical and business objectives on the one hand, and process paralysis on the other. Chapter 8 provides tailoring guidelines to help achieve that balance.

Approved for SEHv3:

Terje Fossnes, CSEP, Chair, INCOSE SEHv3 Development Team

Kevin Forsberg, CSEP, Co-Chair, INCOSE SEHv3 Development Team

Eric Aslaksen, CSEP, INCOSE Associate Director, Technical Review

Samantha Brown, INCOSE Technical Director

Approved for SEHv3.1:

Kevin Forsberg, CSEP, Chair, INCOSE SEHv3.1 Development Team

Erik Aslaksen, CSEP, INCOSE Associate Director, Technical Review

Samantha Brown, INCOSE Technical Director

Approved for SEHv3.2 / v3.2.1:

Kevin Forsberg, ESEP, Chair, INCOSE Knowledge Management Working Group

Yoshi Ohkami, ESEP, INCOSE Associate Director, Technical Review

G. Richard Freeman, CSEP, INCOSE Assistant Director, Technical Information

Regina Griego, INCOSE Technical Director (2010)

Jean-Claude Roussel, INCOSE Technical Director (2011)

This page intentionally left blank.

Table of Contents

Table of Figures

Table of Tables

1 Systems Engineering Handbook Scope

1.1 Purpose

This handbook defines the discipline and practice of systems engineering (SE) for student and practicing professional alike. This handbook provides an authoritative reference to understand the discipline in terms of content and practice.

1.2 Application

This handbook is consistent with ISO/IEC 15288:2008 – *Systems and software engineering – System life cycle processes* (hereafter referred to as ISO/IEC 15288:2008) to ensure its usefulness across a wide range of application domains – man-made systems and products, as well as business and services.

ISO/IEC 15288:2008 is an international standard that is a generic process description, whereas this handbook further elaborates the processes and activities to execute the processes. Before applying this handbook in a given organization or project, it is recommended that the tailoring guidelines in Chapter 8 be used to remove conflicts with existing policies, procedures and standards already in use. Processes and activities in this handbook do not supersede any international, national, or local laws or regulations.

For organizations (including much of commercial industry) that do not follow the principles of ISO/IEC 15288:2008 to specify their life-cycle processes, this handbook can serve as a reference to practices and methods that have proven beneficial to the SE community at large and that can add significant value in new domains if appropriately selected and applied.

1.3 Contents

This chapter defines the purpose and scope of this handbook. Chapter 2 provides an overview of the goals and value of using SE throughout the systems life cycle. Chapter 3 describes an informative life-cycle model with six stages: Concept, Development, Production, Utilization, Support, and Retirement.

ISO/IEC 15288:2008 identifies four process groups to support SE. Each of these process groups is the subject of a chapter. A graphical overview of these processes is given in Figure 1-1.

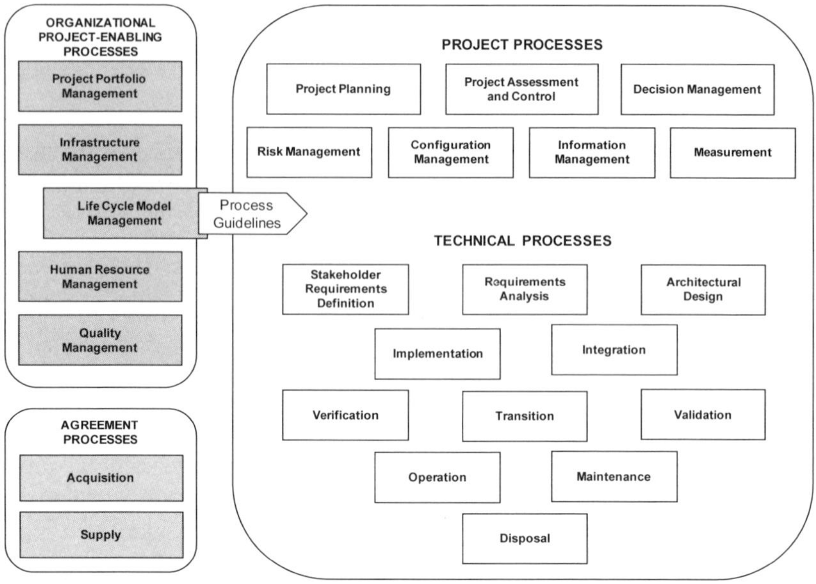

Figure 1-1 System Life-cycle Processes Overview per ISO/IEC 15288:2008

Technical Processes (Chapter 4) include stakeholder requirements definition, requirements analysis, architectural design, implementation, integration, verification, transition, validation, operation, maintenance, and disposal.

Project Processes (Chapter 5) include project planning, project assessment and control, decision management, risk management, configuration management, information management, and measurement.

Agreement Processes (Chapter 6) address acquisition and supply.

Organizational Project-Enabling Processes (Chapter 7) include life-cycle model management, infrastructure management, project portfolio management, human resource management, and quality management**.** As Figure 1-1 illustrates, the outputs of the life-cycle model management process direct the application of the Technical and Project Processes.

NOTE: ISO/IEC 15288:2002 used *Enterprise Processes* that performed the same role as the *Organization Project-Enabling Processes* of ISO/IEC 15288:2008. For the purposes of this v3.2.1 Handbook, the terms Organization and Organizational are used to be synonymous with Enterprise.

Tailoring Processes (Chapter 8). Not every process will apply universally. Careful selection from the material that follows is recommended. Reliance on process over progress will not deliver a system. If you are not familiar with tailoring concepts, please read Chapter 8 before using this handbook.

Specialty Engineering Activities (Chapter 9) contains practical information about topics such as acquisition logistics and human factors engineering (HFE).

Appendix A contains an N^2 analysis of the processes showing where dependencies exist in the form of shared inputs or outputs. Appendix B contains mappings to other international and industry SE standards. Appendices C and D provide a glossary of terms and acronyms. Appendix E contains the acknowledgements for the various versions of this handbook. Errors, omissions, and other suggestions for this handbook can be submitted to the International Council on Systems Engineering (INCOSE) using the Comment Form contained in Appendix F.

1.4 Format

A common format has been applied in Chapters 4 through 8 to the elaboration of the system life-cycle processes found in ISO/IEC 15288:2008. Each process is illustrated by a context diagram. A sample is shown in Figure 1-2. To understand a given process, the reader is encouraged to find the complete information in the combination of diagrams and text. The following heading structure provides consistency in the discussion of these processes.

1. Process Overview
   - Purpose
   - Description
   - Inputs – including Controls and Enablers

     Note: There is often confusion between what is considered an input and what is considered a control. Inputs only include those items that are *acted upon by the process and used or transformed to create the outputs*. For example, guidance and instructions are not inputs. They are considered as part of the controls, since they are not changed by the process.
   - Outputs
   - Process Activities
   - Common approaches and tips

2. Process Elaboration
   - Detail on topics related to the process
3. References.

To ensure consistency with ISO/IEC 15288:2008, the purpose statements from the standard are included for each process. Also, the activities of the standard are included in the context diagrams and in the Process Activities section of each process. This handbook provides additional levels of detail below the activity level.

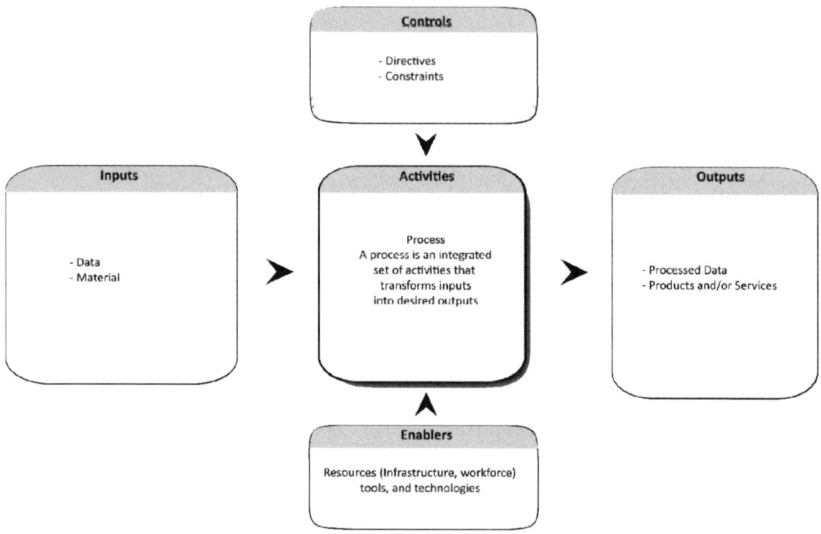

Figure 1-2 Sample of Context Diagram for Process

1.5 Definitions of Frequently Used Terms

One of the Systems Engineer's first jobs on a project is to establish nomenclature and terminology that support clear, unambiguous communication and definition of the system and its functions, elements, operations, and associated processes. While there is no universally accepted nomenclature, this section presents one convention in widespread use.

It is essential to the advancement of the field of SE that common definitions and understandings be established regarding general methods and terminology that in-turn support common processes. As more Systems Engineers accept and use a common terminology, we will experience improvements in communications, understanding, and ultimately, productivity. Table 1-1 contains the definition of some terms used frequently throughout this handbook. Definitions in italics have been taken from ISO/IEC 15288:2008. A full glossary of definitions is found in Appendix D, and an elaboration of the system hierarchy is given in Section 2.4.

Table 1-1 Frequently Used Terms

Term	Definition
Activity	*set of cohesive tasks of a process*
Element	*See System element*
Enabling system	*a system that supports a system-of-interest during its life-cycle stages but does not necessarily contribute directly to its function during operation*
Enterprise	*See Organization*
Organization	*person or a group of people and facilities with an arrangement of responsibilities, authorities and relationships [adapted from ISO 9000:2005]*
Process	*set of interrelated or interacting activities which transforms inputs into outputs*
Project	*an endeavor with start and finish criteria undertaken to create a product or service in accordance with specified resources and requirements*
Stage	*a period within the life cycle of an entity that relates to the state of its description or realization*
System	*a combination of interacting elements organized to achieve one or more stated purposes* an integrated set of elements, subsystems, or assemblies that accomplish a defined objective. These elements include products (hardware, software, firmware), processes, people, information, techniques, facilities, services, and other support elements. (INCOSE) An example would be an air transportation system.
System element	*a member of a set of elements that constitutes a system* a major product, service, or facility of the system, e.g. the aircraft element of an air transportation system (the term subsystem is sometimes used instead of element)
System-of-interest	*the system whose life cycle is under consideration*

Term	Definition
System-of-systems	System of systems applies to a system-of-interest whose system elements are themselves systems; typically these entail large scale inter-disciplinary problems with multiple, heterogeneous, distributed systems.
Systems Engineering	Systems Engineering (SE) is an interdisciplinary approach and means to enable the realization of successful systems. It focuses on defining customer needs and required functionality early in the development cycle, documenting requirements, and then proceeding with design synthesis and system validation while considering the complete problem: operations, cost and schedule, performance, training and support, test, manufacturing, and disposal. SE considers both the business and the technical needs of all customers with the goal of providing a quality product that meets the user needs. (INCOSE)

1.6 References

The following documents have been used to establish the framework and practical foundations for this handbook.

1. ISO/IEC 15288:2008, *Systems and software engineering – System life cycle processes*, Geneva: International Organization for Standardization, issued 1 February 2008.
2. ISO/IEC TR 19760:2003, *Systems Engineering – A guide for the application of ISO/IEC 15288,* Geneva: International Organization for Standardization, issued 15 November 2003.
3. *SE Guidebook for ITS,* California Department of Transportation, Division of Research and Innovation, v 2.0, Jan 2007.

2 Systems Engineering Overview

2.1 Introduction

This chapter offers a brief overview of the discipline of SE, beginning with some definitions, an abbreviated survey of the origins of the discipline, and discussions of the value of applying SE. Systems are pervasive in our daily life. They are tangible in that they exist in the products we use, the technologies we employ, the services we procure, and in the fabric of society.

2.2 Definition of Systems Engineering

Systems engineering is a perspective, a process, and a profession, as illustrated by these three representative definitions.

> *Systems engineering is a discipline that concentrates on the design and application of the whole (system) as distinct from the parts. It involves looking at a problem in its entirety, taking into account all the facets and all the variables and relating the social to the technical aspect.* (Ramo[1])

> *Systems engineering is an iterative process of top-down synthesis, development, and operation of a real-world system that satisfies, in a near optimal manner, the full range of requirements for the system.* (Eisner[2])

> *Systems engineering is an interdisciplinary approach and means to enable the realization of successful systems. It focuses on defining customer needs and required functionality early in the development cycle, documenting requirements, and then proceeding with design synthesis and system validation while considering the complete problem: operations, cost and schedule, performance, training and support, test, manufacturing, and disposal. SE considers both the business and the technical needs of all customers with the goal of providing a quality product that meets the user needs.* (INCOSE[3])

Certain keywords emerge from this sampling – interdisciplinary, iterative, socio-technical, and wholeness.

The SE perspective is based on systems thinking. Systems thinking occurs through discovery, learning, diagnosis, and dialog that lead to sensing, modeling, and talking about the real-world to better understand, define, and work with systems. Systems thinking is a unique perspective on reality—a perspective that sharpens our awareness of wholes and how the parts within those wholes interrelate. A systems thinker knows how systems fit into the larger context of day-to-day life, how they behave, and how to manage them.

Systems thinking recognizes circular causation, where a variable is both the cause and the effect of another and recognizes the primacy of interrelationships and non-linear and organic thinking—a way of thinking where the primacy of the whole is acknowledged.

The SE process has an iterative nature that supports learning and continuous improvement. As the processes unfold, systems engineers uncover the real requirements and the emergent properties of the system. Complexity can lead to unexpected and unpredictable behavior of systems; hence, one of the objectives is to minimize undesirable consequences. This can be accomplished through the inclusion of and contributions from experts across relevant disciplines coordinated by the systems engineer.

Since SE has a horizontal orientation, the discipline (profession) includes both technical and management processes. Both processes depend upon good decision making. Decisions made early in the life cycle of a system, whose consequences are not clearly understood, can have enormous implications later in the life of a system. It is the task of the systems engineer to explore these issues and make the critical decisions in a timely manner. The role of the systems engineer is varied, and Sheard's "Twelve Systems Engineering Roles"[4] provides one description of these variations.

2.3 Origins of Systems Engineering

The modern origins of SE can be traced to the 1930's followed quickly by other programs and supporters.[5] Table 2-1 and Table 2-2 offer a thumbnail of some important highlights in the origins and history of the application of SE.

Table 2-1 Important Dates in the Origins of SE as a Discipline

1829	Rocket locomotive; progenitor of main-line railway motive power
1937	British multi-disciplinary team to analyze the air defense system
1939-1945	Bell Labs supported NIKE development
1951-1980	SAGE Air Defense System defined and managed by MIT
1956	Invention of systems analysis by RAND Corporation
1962	Publication of *A Methodology for Systems Engineering*
1969	Jay Forrester (Modeling Urban Systems at MIT)
1990	NCOSE established
1995	INCOSE emerged from NCOSE to incorporate International view

Table 2-2 Evolution of SE Standards[6]

1969	Mil-Std 499
1974	Mil-Std 499A
1979	Army Field Manual 770-78
1994	Mil-Std 499B (not released)
1994	Perry Memorandum urges military contractors to adopt commercial practices. EIA 632 IS (Interim Standard) and IEEE 1220 (Trial Version) instead of Mil-Std 499B
1998	EIA 632 Released
1999	IEEE 1220 Released
2002	Release of ISO/IEC 15288:2002
2008	Release of ISO/IEC 15288:2008

With the introduction of the international standard ISO/IEC 15288 in 2002, the discipline of SE was formally recognized as a preferred mechanism to establish agreement for the creation of products and services to be traded between two or more organizations – the supplier(s) and the acquirer(s). But even this simple designation is often confused in a web of contractors and subcontractors since the context of most systems today is as a part of a "system of systems" (see Section 2.5).

2.4 The Hierarchy *Within* A System

The systems considered in ISO/IEC 15288:2008:

> *are man-made, created and utilized to provide products and/or services in defined environments for the benefit of users and other stakeholders. These systems may be configured with one or more of the following system elements: hardware, software, data, humans, processes (e.g., processes for providing services to others), procedures (e.g., operator instructions), facilities, materials and naturally occurring entities. In practice, they are thought of as products or services.*

> *The perception and definition of a particular system, its architecture and its system elements depend on an observer's interests and responsibilities. One person's system-of-interest can be viewed as a system element in another person's system-of-interest. Furthermore, a system-of-interest can be viewed as being part of the environment of operation for another person's system-of-interest.*[7]

Hierarchies are system or organizational representations of a partitioning relationship where the entity (or system) is separated into smaller more manageable entities. The hierarchy should be balanced with appropriate fan-out and span of control. Appropriate fan-out and span of control refers to the number of elements subordinate to each element in the hierarchy. The concept of a system hierarchy or system structure is described in ISO/IEC 15288:2008 as follows:

> *The system life cycle processes ... are described in relation to a system that is composed of a set of interacting system elements, each of which can be implemented to fulfill its respective system requirements. Responsibility for the implementation of any system element may therefore be delegated to another party through an agreement.*
>
> *The relationship between the system and its complete set of system elements can typically be represented in a two-level hierarchy for the simplest of systems-of-interest. For more complex systems-of-interest, a prospective system element may itself need to be considered as a system (that in turn is comprised of system elements) before a complete set of system elements can be defined with confidence (see Figure 2-1). In this manner, the appropriate system life cycle processes are applied recursively to a system-of-interest to resolve its structure to the point where understandable and manageable system elements can be implemented (made, bought, or reused) from another party. The system-of-interest may include any type of system or combination of systems.*[8]

System hierarchies are analogous to organizational hierarchies in that both can suffer from improper balance; that is, too great a span of control or excessive layers in the hierarchy. A "rule of thumb" useful in evaluating this balance is that a system should have no more than 7 ± 2 elements reporting to it. In the same way an element should have no more than 7 ± 2 subsystems reporting to it, and so on. A design level with too many subordinate entities suffers from too much complexity. The design and corresponding verification activities run the risk of running out-of-control or acquiring an informal partitioning that guides the work without proper control or visibility. A level of design with too few subordinate entities likely does not have distinct design activity, and both design and verification activities contain redundancy.

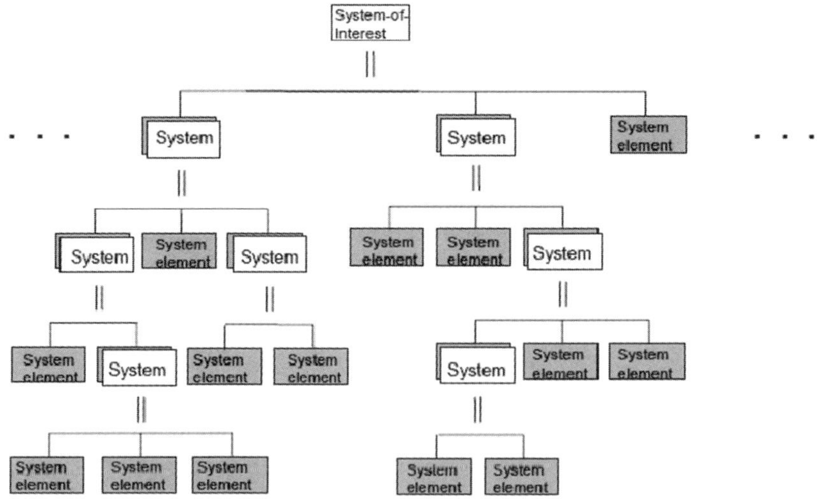

Figure 2-1 Hierarchy Within a System[9]

The depth and nomenclature of the hierarchy can and should be adjusted to fit the complexity of the system. Systems elements, for example, are commonly referred to as sub-systems, assemblies, components, parts, etc. at progressively lower-levels of the system hierarchy. In the complex Apollo program, NASA added a "Module Level" in the hierarchy to breakout the Command Module, Lunar Module, etc. of the Space Vehicle Element. In Information Systems, elements might include computers, networks, printers, data storage, and personnel. Simple systems typically have fewer levels in the hierarchy than complex systems.

2.5 Systems of Systems

"Systems-of-Systems" (SoS) are systems-of-interest whose system elements are themselves systems; typically these entail large-scale inter-disciplinary problems involving multiple, heterogeneous, distributed systems. These interoperating collections of component systems usually produce results unachievable by the individual systems alone.[10]

Figure 2-2 illustrates these concepts. The Global Positioning System (GPS), which is an integral part of the navigation system on board an aircraft, is a system in its own right rivaling the complexity of the air transportation system. Another characteristic of SoS is that the component systems may be part of other unrelated systems. For instance, the GPS may be an integral part of automobile navigation systems.

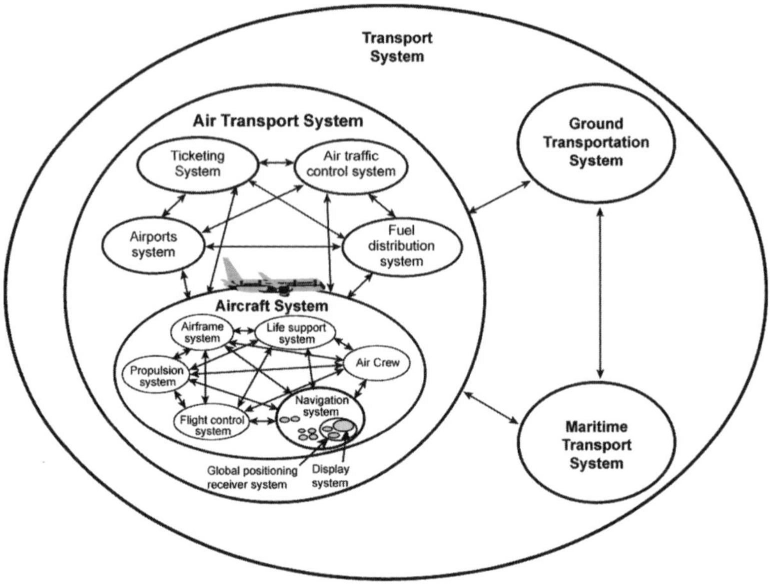

Figure 2-2 Example of the multitude of perceivable systems of interest in an aircraft and its environment of operation within a Transport system of Systems[11]

Not all SoS involve an environment as complex as the air transportation system. A digital camera may seem simple, but it is an SoS with rigidly controlled interfaces. Multiple camera bodies, from simple fixed focus digital cameras to sophisticated single lens reflex cameras, have a common interface to digital memory cards. The full single-lens reflex camera system has many different models of camera bodies, which interface with 50 or more lens systems and multiple flash units. To be a commercial success, these simple to sophisticated camera systems are designed to conform to external interfaces for standard commercial batteries, compact flash memory cards, interface cables, computers, and printer software, as illustrated in Figure 2-3. In the context of an SoS, systems are enclosed in the white boxes, system elements are displayed in the gray area.

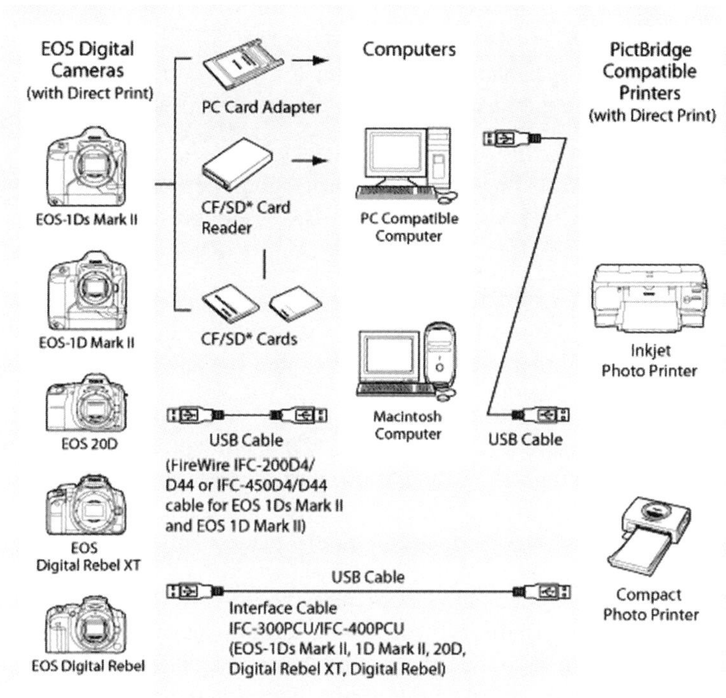

Figure 2-3 Digital Camera and Printer System of Systems[12]

The following challenges all influence the development of systems of systems:

1. *System elements operate independently.* – Each system in an SoS is likely to be operational in its own right.

2. *System elements have different life cycles.* – An SoS involves more than one system element. Some of the system elements are possibly in their development life cycle while others are already deployed as operational. In extreme cases, older systems elements in an SoS might be scheduled for disposal before newer system elements are deployed.

3. *The initial requirements are likely to be ambiguous.* – The requirements for an SoS can be very explicit for deployed system elements. But for system elements that are still in the Development Stage, the requirements are usually no more explicit than the system element requirements. Requirements for an SoS mature as the system elements mature.

4. *Complexity is a major issue.* – As system elements are added, the complexity of system interaction grows in a non-linear fashion. Furthermore, conflicting or missing interface standards can make it hard to define data exchanges across system element interfaces.
5. *Management can overshadow engineering.* – Since each system element has its own product/project office, the coordination of requirements, budget constraints, schedules, interfaces, and technology upgrades further complicate the development of an SoS.
6. *Fuzzy boundaries cause confusion.* – Unless someone defines and controls the scope of an SoS and manages the boundaries of system elements, no one controls the definition of the external interfaces.
7. *SoS engineering is never finished.* – Even after all system elements of an SoS are deployed, product/project management must continue to account for changes in the various system element life cycles, such as new technologies that impact one or more system elements, and normal system replacement due to pre-planned product improvement.

Part of the systems engineer's job in an SoS environment is to be aware of and mitigate the risk of each of these seven challenges. Focus is placed on controlling the interfaces between system elements and external systems. It is especially important to ensure that the interfaces are still operational when an older component system is replaced with a newer version. Verification and validation (V&V) processes play a critical role in such transitions.

2.6 Use of Systems Engineering

It can be readily inferred from the nature of the earliest projects that the SE discipline emerged as an effective way to manage complexity and change. As both complexity and change continue to escalate in our products, services, and society, reducing the risk associated with new systems or modifications to complex systems continues to be a primary goal of the systems engineer. This is illustrated in Figure 2-4. The percentages along the time line represent the actual life-cycle cost (LCC) accrued over time based on a statistical analysis performed on projects in the U.S. Department of Defense (DOD) as reported by the Defense Acquisition University. As shown, the Concept Stage of a new system averages 8% of the total LCC. The curve for committed costs represents the amount of LCC committed by project decisions and indicates that when 20% of the actual cost has been accrued, 80% of the total LCC has already been determined. The diagonal arrow under the curve reminds us that errors are less expensive to remove early in the life cycle.

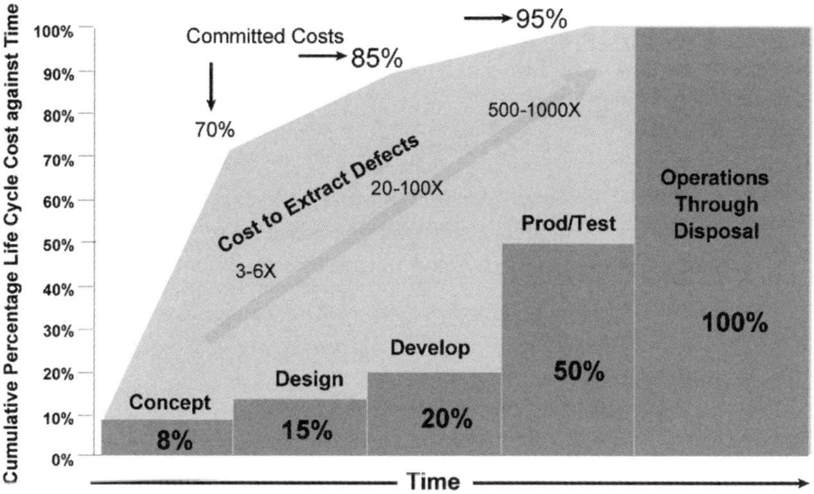

Figure 2-4 Committed Life-cycle Cost against Time[13]

Figure 2-4 also demonstrates the consequences of making early decisions without the benefit of good information and analysis. SE extends the effort performed in concept exploration and design to exceed the percentages shown in the cumulative committed cost curve and reduce the risk of hasty commitments without adequate study. Though shown as linear, the execution of the various life-cycle stages associated with modern product development is, in actual application, recursive. Nonetheless, the consequences of ill-formed decisions at throughout the life cycle are the same.

Another factor driving the need for SE is that the time from prototype to significant market penetration of a new product has dropped by more than a factor of four in the past 50 years (see Figure 2-5). Complexity has an impact on innovation. Few new products represent the big-bang introduction of new invention; rather, most products and services in today's market are the result of incremental improvement. This means that the life cycle of today's products and services is longer and subject to increasing uncertainty. A well-defined SE process becomes critical to establishing and maintaining a competitive edge in the 21st century.

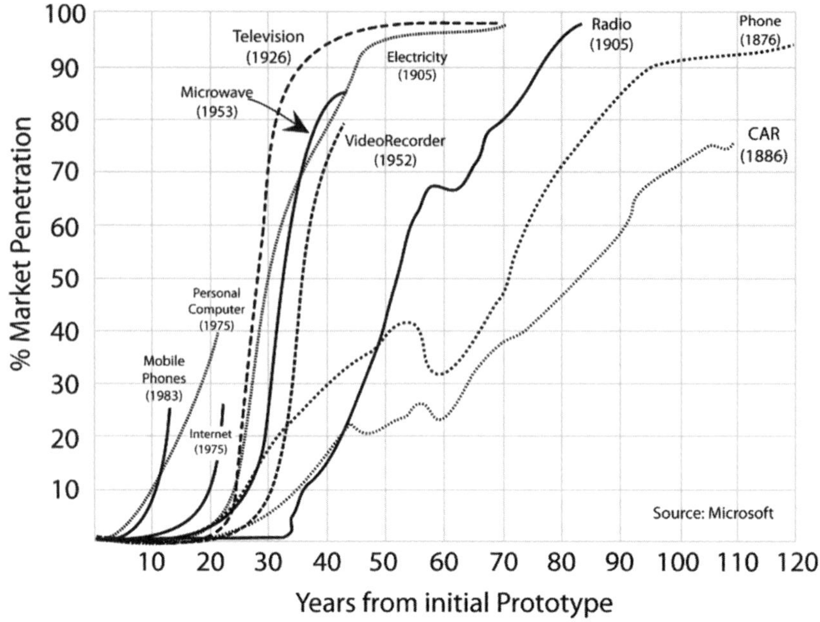

Figure 2-5 In the last century, the time from prototype to significant market penetration is dramatically reduced[14]

The message is that the development of technology has accelerated over the past 100 years. This is illustrated in summary Figure 2-6 below. In the last century, in this sample of products, the time it took to achieve 25% market penetration was reduced from about 50 to below 12 years. On average, development from prototype to 25% market penetration went from 44 years to 17 years.

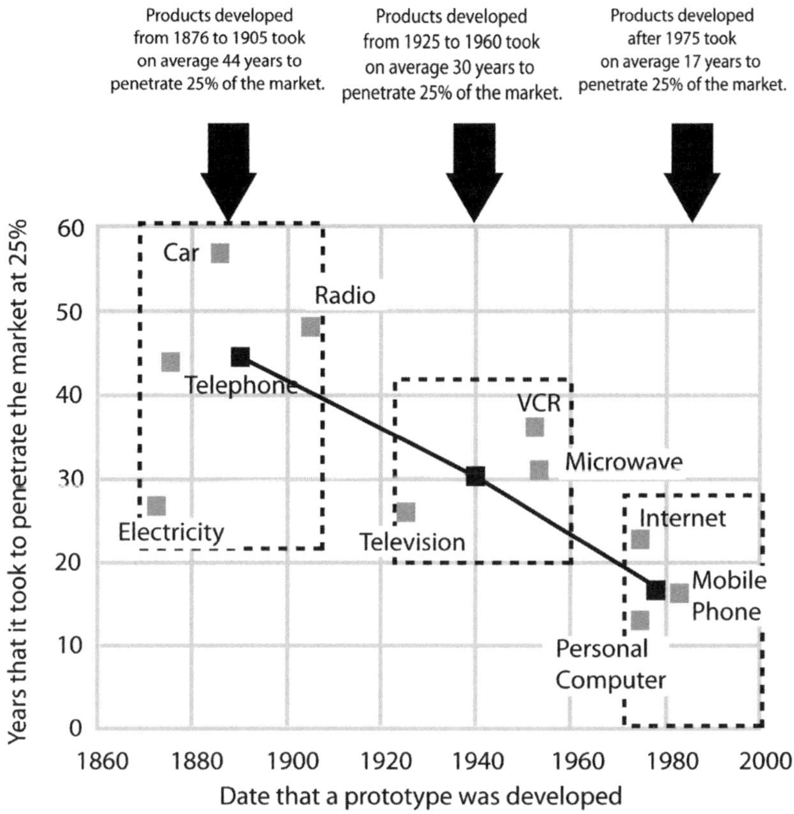

Figure 2-6 Technology acceleration over the past 140 years

2.7 Value of Systems Engineering

A study researching the return on investment from using SE was conducted by the INCOSE Systems Engineering Center of Excellence (SECOE) beginning in 2001. The results uncovered an inverse correlation between cost and schedule overruns and the amount of SE effort applied to a project or development activity.

Cost and schedule overruns on the reported projects are illustrated in Figure 2-7. The following effects are apparent:

Cost

- Cost overrun lessens with increasing SE effort and appears to minimize at something greater than 10% SE effort.

- Variance in the cost overrun also lessens with increasing SE effort. At low SE effort, a project has difficulty predicting its overrun, which may be between 0% (actual = planned) and 200% (actual = 3 x planned). At 12% SE effort, the project cost is more predictable, falling between minus 20% (actual = 0.80 x planned) and 41% (actual = 1.41 x planned). The dashed lines are the 90th percentile when assuming a normal distribution.

Schedule

- Schedule overrun lessens with increasing SE effort and appears to minimize at something greater than 10% SE effort, although few data points exist to support a reliable calculation. The solid line is the least-squares trend line for a second order curve.

- Variance in the schedule overrun also lessens with increasing SE effort. At low SE effort, a project has difficulty predicting its overrun, which may be between minus 35% (actual = 0.65 x planned) and 300% (actual = 4 x planned). At 12% SE effort, the project schedule is more predictable, falling between minus 22% (actual = 0.78 x planned) and 22% (actual = 1.22 x planned). The dashed lines are the 90th percentile when assuming a normal distribution.

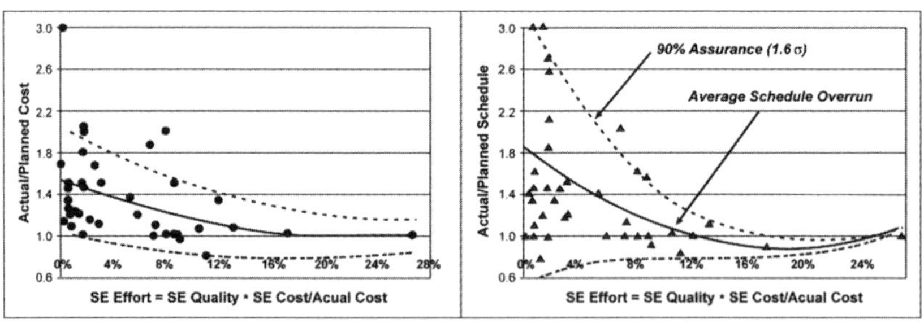

Figure 2-7 Cost and schedule overruns correlated with SE effort[15]

Additional work is underway to collect more data about the value of applying SE to a project. These initial results indicate that SE effort can be a positive factor in controlling cost and schedule overruns and reducing the uncertainty of project execution.

2.8 An Allegorical Tale

A senior systems engineer at a major U.S. company visited all of the divisions with the goal of increasing the use of good system engineering practices. His message included all the things that SE can/should do in commercializing products. His message also included a strong bias towards planning and documentation. Over a period of months he visited with Division Managers, Chief Engineers, Program Managers, and Senior Engineers. He returned completely depleted of his enthusiasm. The problem was that the message was totally rejected because it was interpreted as useless, non-value added work or way beyond anything the company could afford to do from a time and dollars perspective. Sometime later, another senior systems engineer visited many of the same people with the same purpose but a different message. The message the second engineer delivered was that big gains could be made by focusing on the most important customer needs and using a select group of synergistic system engineering tools/practices to fill those needs. This time the message was well received.

The lesson: "Systems engineering is a multi-disciplinary effort that involves both the technical effort and technical project management aspects of a project. Organizations seeking to incorporate the benefits of processes outlined in ISO/IEC 15288:2008 will remember that application of those processes, and the enablers discussed in this handbook, requires vision and practical application of the principles."[16]

2.9 References

1. Federal Aviation Administration [USA], *Systems Engineering Manual,* definition contributed by Simon Ramo
2. Eisner, Howard, *Essentials of Project and Systems Engineering Management*
3. INCOSE, *What is Systems Engineering?*, http://www.incose.org/practice/whatissystemseng.aspx, 14 June 2004
4. Sheard, Sarah, "Twelve Roles of SE," *Proceedings of the 6th Annual INCOSE International Symposium*, 1996
5. Hughes, Thomas P., *Rescuing Prometheus*, Chapter 4, pp. 141-195, Pantheon Books, New York, 1998
6. Courtesy of James Martin
7. ISO/IEC 15288:2008, p. 7-8
8. Ibid, p. 8-9
9. Ibid, p. 9, Figure 2
10. Krygiel, Annette J., *Behind the Wizard's Curtain*, CCRP Publication Series, July 1999, p 33
11. ISO/IEC 15288:2002, p. 53, Figure D-1
12. Canon EOS Digital Camera Brochure
13. Defense Acquisition University, 1993
14. Microsoft
15. Honour, Eric, (2004), "Understanding the Value of Systems Engineering," *Proceedings of the 14th Annual INCOSE International Symposium*, 1996, available online from the INCOSE Systems Engineering Center of Excellence (SECOE), <http://www.incose.org/secoe>
16. Submitted by the *INCOSE Systems Engineering Handbook* v3.1 review team

This page intentionally left blank.

3 Generic Life-Cycle Stages

3.1 Introduction

Every man-made system has a life cycle, even if it is not formally defined. In keeping with increased awareness of environmental issues, the life cycle for any system-of-interest must encompass not only the development, production, and Utilization Stages but also provide early focus on the Retirement Stage when decommissioning and disposal of the system will occur.

The role of the systems engineer encompasses the entire life cycle for the system-of-interest. Systems engineers orchestrate the development of a solution from requirements determination through operations and system retirement by assuring that domain experts are properly involved, that all advantageous opportunities are pursued, and that all significant risks are identified and mitigated. The systems engineer works closely with the project manager in tailoring the generic life cycle, including key decision gates, to meet the needs of their specific project. Per ISO/IEC 15288:2008:

> *5.2.2 – Life cycles vary according to the nature, purpose, use and prevailing circumstances of the system. Each stage has a distinct purpose and contribution to the whole life cycle and is conserved when planning and executing the system life cycle. ... The stages thus provide organizations with a framework within which organization management has high-level visibility and control of project and Technical Processes.*[1]

> *6.2.1.3 (a)(5) NOTE – The life cycle processes and activities are selected, tailored as appropriate and employed in a stage to fulfill the purpose and outcomes of that stage.*[2]

The purpose in defining the system life cycle is to establish a framework for meeting the stakeholders' needs in an orderly and efficient manner. This is usually done by defining life-cycle stages and using decision gates to determine readiness to move from one stage to the next. Skipping stages and eliminating "time consuming" decision gates can greatly increase the risks (cost and schedule) and may adversely affect the technical development as well by reducing the level of the SE effort, as discussed in Section 2.6.

Systems engineering tasks are usually concentrated at the beginning of the life cycle, but both commercial and government organizations recognize the need for SE throughout the systems life span, often to modify or change a system product or service after it enters production or is placed in operation. Subsequently, SE is an important part of all life cycle stages. During Operations and Support (O&S) stages, for example, SE executes performance analysis, interface monitoring, failure analysis, logistics analysis, tracking, and management, etc. that is essential to ongoing support of the system.

3.2 Life Cycle Characteristics

3.2.1 Three Aspects of the Life Cycle

Every system life cycle consists of the business aspect (business case), the budget aspect (funding), and the technical aspect (product). The systems engineer creates technical solutions that are consistent with the business case and the funding constraints. System integrity requires that these three aspects are in balance and given equal emphasis at all decision gate reviews. For example, when Motorola's Iridium project started in the late 1980s, the concept of satellite-based mobile phones was a breakthrough and would clearly capture a significant market share. Over the next dozen years, the technical reviews ensured a highly successful technical solution. In fact, in the first decade of the 21st century, the Iridium project is proving to be a good business venture for all except for the original team who had to sell all the assets—at about two percent of their investment—through the bankruptcy court. The original team lost sight of the competition and changing consumer patterns that substantially altered the original business case. Figure 3-1 highlights two critical parameters that engineers sometimes lose sight of: time to break even (indicated by the circle) and Return on Investment (indicated by the lower curve).

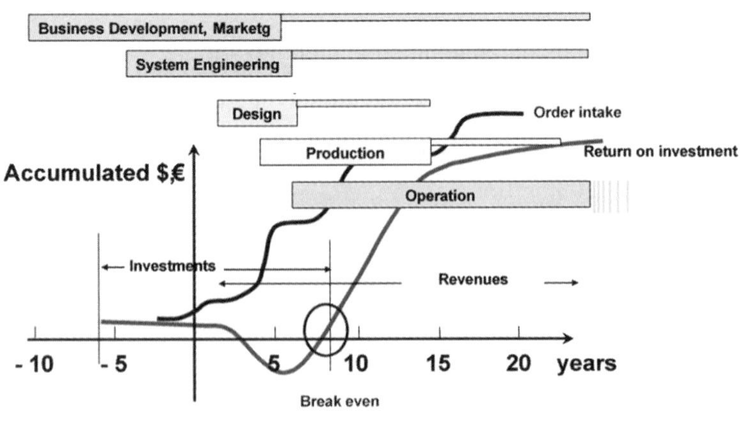

Figure 3-1 Generic business Life-cycle[3]

3.2.2 Decision Gates

Decision gates, also known as control gates, are often called "Milestones" or "Reviews." A decision gate is an approval event in the project cycle, sufficiently important to be defined and included in the schedule by the project manager, executive management, or the customer. Entry and exit criteria are established for each gate at the time they are included into the project management baseline. Decision gates ensure that new activities are not pursued until the previously scheduled activities, on which new activities depend, are satisfactorily completed and placed under configuration control. Proceeding beyond the Decision Gate before the project is ready entails risk. The project manager may decide to accept that risk, as is done, for instance, with long-lead item procurement.

All decision gates are both reviews and milestones; however, not all reviews and milestones are decision gates. Decision gates address the following questions:

- Does the project deliverable still satisfy the business case?
- Is it affordable?
- Can it be delivered when needed?

Decision gates represent major decision points in the system life cycle. The primary objectives of decision gates are to:

- Ensure that the elaboration of the business and technical baselines are acceptable and will lead to satisfactory V&V
- Ensure that the next step is achievable and the risk of proceeding is acceptable
- Continue to foster buyer and seller teamwork
- Synchronize project activities.

There are at least two decision gates in any project: authority to proceed and final acceptance of the project deliverable. The project team needs to decide which life-cycle stages are appropriate for their project and which decision gates beyond the basic two are needed. Each decision must have a beneficial purpose; "pro-forma" reviews waste everyone's time. Even in Agile Development frequent interaction with the customer may minimize, but not eliminate, the need for decision gates. The consequences of conducting a superficial review, omitting a critical discipline, or skipping a decision gate altogether are usually long-term and costly.

The project business case issues of market demand, affordability, and realistic schedules are important decision criteria influencing concept selection, and they should be updated and evaluated at every decision gate. Inadequate checks along the way can set up subsequent failures – usually a major factor in cost over-runs and delays. At each gate the decision options are:

- *Acceptable* – Proceed with the next stage of the project
- *Acceptable with reservations* – Proceed and respond to action items
- *Unacceptable: Do not proceed* – Continue this stage and repeat the review when ready
- *Unacceptable: Return to a preceding stage*
- *Unacceptable: Put a hold on project activity*
- *Unsalvageable: Terminate the project.*

Upon successful completion of a decision gate, some artifacts (e.g., documents, models, or other products of a project cycle stage) have been approved as the basis upon which future work must build. If the project is large or long enough, or entails high risk, these artifacts are placed under configuration management.

Decision gate descriptions should identify the:

- Purpose of the decision gate
- Host and chairperson
- Attendees
- Location
- Agenda and how the decision gate is to be conducted
- Evidence to be evaluated
- Actions resulting from the decision gate
- Method of closing the review.

Decision gate approval follows review by qualified experts and involved stakeholders and is based on hard evidence of compliance to the criteria of the review. One of the underlying principles for the Agile Development and extreme programming movements is to substantially reduce (but not eliminate) the frequency and elaborate (and they would claim pro-forma) content of decision gates for software development. Balancing the formality and frequency of decision gates is seen as a critical success factor for all SE process areas. On large or lengthy projects, decisions and their rationale are maintained using an Information Management Process.

3.3 Life-Cycle Stages

ISO/IEC 15288:2008 states:

> *5.2.1 – Every system has a life cycle. ... A system progresses through its life cycle as the result of actions, performed and managed by people in organizations, using processes for execution of these actions.*[4]

6.2.1.3 (a)(5) NOTE – The life cycle model comprises one or more stage models, as needed. It is assembled as a sequence of stages that may overlap and/or iterate, as appropriate for the system-of-interest's scope, magnitude, complexity, changing needs and opportunities.[5]

The detailed SE processes applied during the life cycle model stages are tailored and expressed in terms of the ISO/IEC 15288:2008 processes and their outcomes, relationships, and sequence. Table 3-1 lists seven generic life-cycle stages. The purpose of each is briefly identified in the table, and the options from decision gates events are indicated. Note that stages can overlap, and the utilization and Support Stages run in parallel. Note also that the outcome possibilities for decision gates are the same for all decision gates, from the first in the concept review to the last in the Retirement Stage.

Table 3-1 Generic life-cycle stages, their purposes, and decision gate options[6]

<table>
<tr><th>LIFE-CYCLE STAGES</th><th>PURPOSE</th><th>DECISION GATES</th></tr>
<tr><td>EXPLORATORY RESEARCH</td><td>Identify stakeholders' needs
Explore ideas and technologies</td><td rowspan="7">Decision Options
– Proceed with next stage
– Proceed and respond to action items
– Continue this stage
– Return to preceding stage
– Put a hold on project activity
– Terminate project.</td></tr>
<tr><td>CONCEPT</td><td>Refine stakeholders' needs
Explore feasible concepts
Propose viable solutions</td></tr>
<tr><td>DEVELOPMENT</td><td>Refine system requirements
Create solution description
Build system
Verify and validate system</td></tr>
<tr><td>PRODUCTION</td><td>Produce systems
Inspect and verify</td></tr>
<tr><td>UTILIZATION</td><td>Operate system to satisfy users' needs</td></tr>
<tr><td>SUPPORT</td><td>Provide sustained system capability</td></tr>
<tr><td>RETIREMENT</td><td>Store, archive, or dispose of the system</td></tr>
</table>

Subsequent chapters of this handbook will define processes and activities to meet the objectives of these life-cycle stages. Because of the iterative nature of SE, specific processes are not aligned to individual life cycle stages. Rather, the entire set of SE processes is considered and applied at each stage of life cycle development as appropriate to the scope and complexity of the project. Figure 3-2 shows the estimated level of effort for SE processes as they are applied across the life cycle of a project.

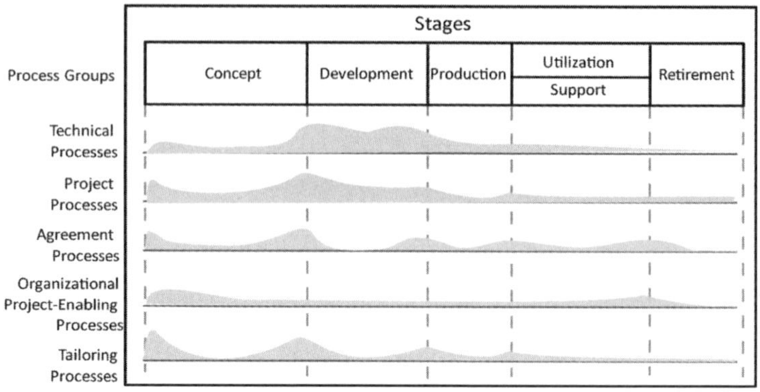

Figure 3-2 SE level of effort across life-cycle stages[7]

Figure 3-3 compares the generic life-cycle stages to other life-cycle viewpoints. For example, the Concept Stage is aligned with the commercial project's Study Period and with the Pre-systems Acquisition and the Project Planning Period in the U.S. Departments of Defense and Energy, respectively. Typical decision gates are presented in the bottom line.

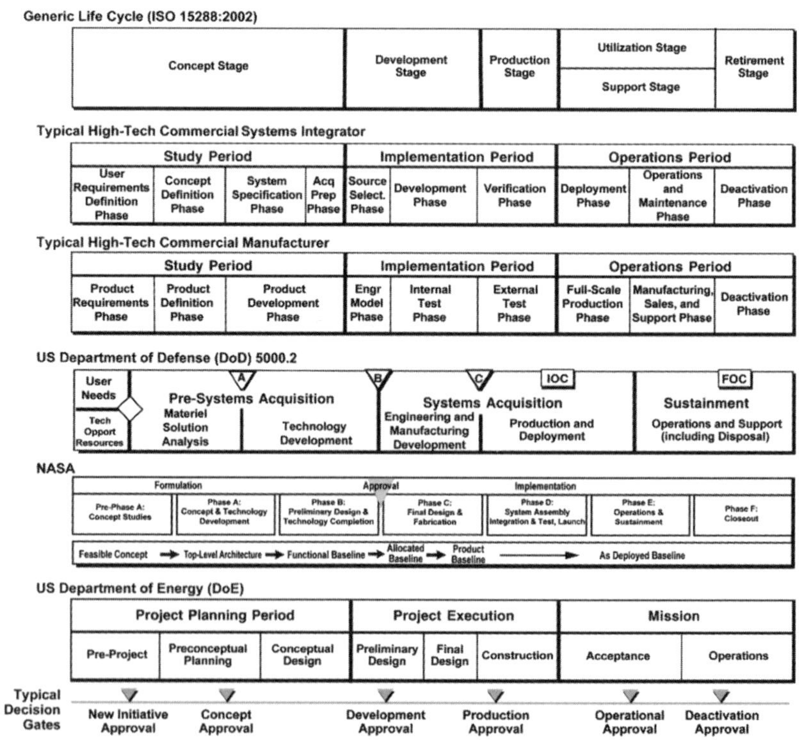

Figure 3-3 Comparisons of life-cycle models[8]

Various life-cycle models, such as the waterfall, spiral, Vee, and Agile Development models, are useful in defining the start, stop, and process activities appropriate to the life-cycle stages. The Vee model (see Figure 3-4) is used to visualize the system engineering focus, particularly during the Concept and Development Stages. The Vee highlights the need to define verification plans during requirements development, the need for continuous validation with the stakeholders, and the importance of continuous risk and opportunity assessment.

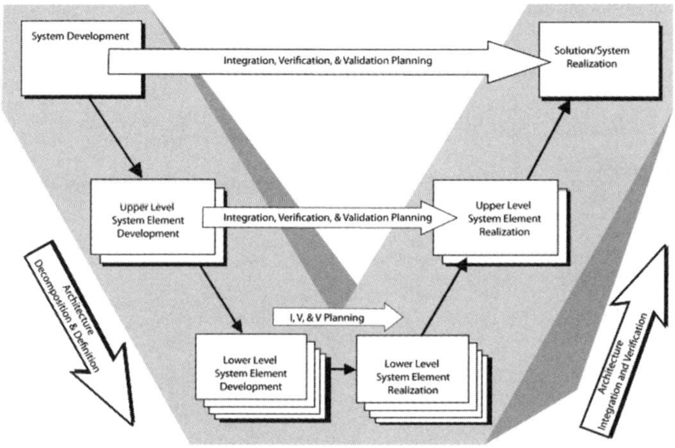

Figure 3-4 Vee model[9]

The Vee model provides a useful illustration of the SE activities during the life-cycle stages. In the Vee model, time and system maturity proceed from left to right. The core of the Vee (i.e., those products that have been placed under configuration control) depicts the evolving baseline from user requirements agreement to identification of a system concept to definition of elements that will comprise the final system. With time moving to the right, the evolving baseline defines the left side of the core of the Vee, as shown in the shaded portion of Figure 3-5.

As entities are constructed, verified and integrated, the right side of the core of the Vee is executed. Since one can never go backward in time, all iterations in the Vee are performed on the vertical "time now" line. Upward iterations involve the stakeholders and are the in-process validation activities that ensure that the proposed baselines are acceptable. The downward vertical iterations are the essential off-core opportunity and risk management investigations and actions. In each stage of the system life cycle, the SE processes iterate to ensure that a concept or design is feasible and that the stakeholders remain supportive of the solution as it evolves.

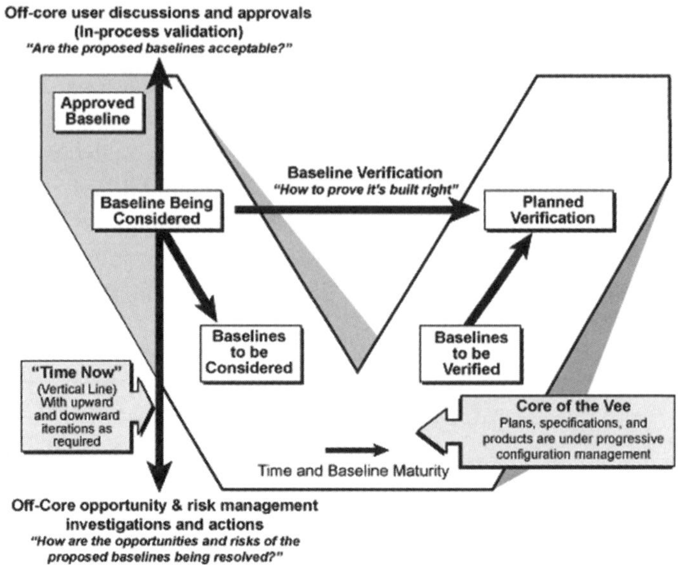

Figure 3-5 Left side of the Vee model[10]

3.3.1 Exploratory Research Stage

Many industries employ an Exploratory Research Stage to study new ideas or enabling technologies and capabilities, which then mature into the initiation of a new project (system-of-interest). A great deal of creative SE is done in this stage, and the systems engineer leading these studies is likely to follow a new idea into the Concept Stage, perhaps as project champion. Often the Exploratory Research activities identify the enabling technologies. As discussed in Chapter 2, if the work is done properly in early stages of the life cycle, it is possible to avoid recalls, and rework in later stages.

Many life cycle models show the process beginning with "Requirements" or "User Requirements." In fact, the process begins earlier with interactions and studies to understand user needs. It is critical that in these early studies a reference architecture and design be created and explored to whatever depth is necessary to identify technological risks and to assess the Technology Readiness Level of the project. In some instances, the project may be an outgrowth of research activities where the research engineer or scientist has no connection to a user-supported need.[11] The reference architecture and enabling technologies need to be identified early, and issues arising from the studies need to be addressed during the system Development Stage.[12]

The reference architecture will also be used to generate early cost and schedule projections for the project if it moves ahead. The first directive from the customer in the Exploratory Research Stage is to clearly identify user needs, and, while avoiding any design work, to provide an estimate of the cost and schedule for the full-scale development. Incomplete SE in this stage can lead to poor cost and schedule projections, which in turn can lead to project support issues as the real project cost and schedule become known during development. For example, the Mars Science Laboratory Rover, scheduled for launch in 2009, had to be "delayed because of technical glitches." This resulted in missing the launch window, causing a two-year delay and a 35% cost growth over the approved development costs. Program critics, however, claimed a 400% cost growth based on the early concept studies, and they threatened the project with cancellation as a result.[13]

The reference architecture is a starting point, not an end point, as the project moves into the Concept Stage (see Section 3.3.2). The reference architecture is not put under configuration control, and the key output from the Exploratory Research Stage is a clearer understanding of the user needs, an assessment of the Technology Readiness to move to the next stage, and a rough estimate of the project cost and schedule requirements to first article delivery.

3.3.2 Concept Stage

The Concept Stage is a refinement and broadening of the studies, experiments, and engineering models pursued during the Exploratory Research Stage. The processes described in this handbook are requirements-driven, as opposed to product-driven. Thus, the first step is to identify, clarify, and document stakeholders' requirements. If there was no Exploratory Research stage, that effort is done here.

During the Concept Stage, the team begins in-depth studies that evaluate multiple candidate concepts and eventually provide a substantiated justification for the system concept that is selected. As part of this evaluation, mockups may be built (for hardware) or coded (for software), engineering models and simulations may be executed, and prototypes of critical elements may be built and tested. Prototypes are helpful to verify the feasibility of concepts, to aid the understanding of user needs, and to explore risks and opportunities. These studies expand the risk and opportunity evaluation to include affordability assessment, environmental impact, failure modes, hazard analysis, technical obsolescence, and system disposal. The systems engineer facilitates these analyses by coordinating the activities of engineers from many disciplines. Key objectives are to provide confidence that the business case is sound and the proposed solutions are achievable.

The Concept Stage includes system, element, and key subsystem-level concept and architecture definition, as well as integration, verification, and validation (IV&V) planning. Early validation efforts align requirements with stakeholder expectations. The systems capabilities specified by the stakeholders will be met by the combination of system elements. Problems identified for individual hardware parts or software modules should be addressed early to minimize the risk that, when these entities are finally designed and verified, they fall short of the required functionality or performance.

Many projects are driven by eager project champions who want "to get on with it." They succumb to the temptation to cut short the Concept Stage, and they use exaggerated projections to support starting detailed design without adequate understanding of the challenges involved, as comically illustrated in Figure 3-6. Many commissions reviewing failed systems after the fact have identified insufficient or superficial study in the Concept Stage as a root cause of failure.

Figure 3-6 Importance of the Concept Stage[14]

3.3.3 Development Stage

The Development Stage includes detailed planning, development, and IV&V activities. Figure 3-7 illustrates the evolving baseline as system elements are integrated and verified. A source of additional information about IV&V and the significance for project cost and risk when these activities are optimized was the subject of the European Union SysTest[14] program.

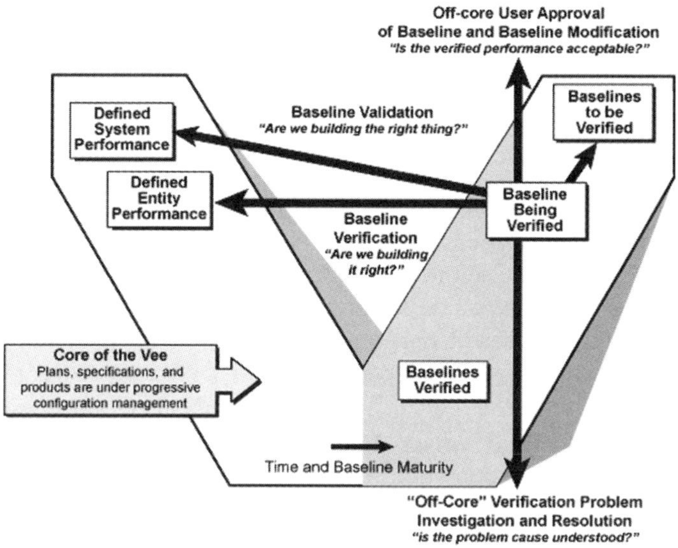

Figure 3-7 Right side of the Vee Model[16]

Per ISO/IEC 15288:2008:

> "5.2.2 – *Organizations employ stages differently to satisfy contrasting business and risk mitigation strategies. Using stages concurrently and in different orders can lead to life cycle forms with distinctly different characteristics.*"[17]

A discussion of system life-cycle stages does not imply that the project should follow a predetermined set of activities or processes unless they add value toward achieving the final goal. Representations of stages tend to be linear in graphical depictions, but this hides the true incremental and iterative nature of the underlying processes. The approaches that follow imply full freedom to choose a development model and are not restricted to a waterfall or other plan-driven methods. For the Development Stage, as for all stages, the organization will select the processes and activities that best suit the project needs.

3.3.4 Production Stage

The Production Stage is where the system-of-interest is produced or manufactured. Product modifications may be required to resolve production problems, to reduce production costs, or to enhance product or system-of-interest capabilities. Any of these may influence system requirements and may require system re-verification or re-validation. All such changes require SE assessment before changes are approved.

3.3.5 Utilization Stage

The Utilization Stage is where the system-of-interest is operated in its intended environment to deliver its intended services. Product modifications are often planned for introduction throughout the operation of the system. Such upgrades enhance the capabilities of the system. These changes should be assessed by systems engineers to ensure smooth integration with the operational system-of-interest. The corresponding technical process is the Operations Process.

3.3.6 Support Stage

The Support Stage is where the system-of-interest is provided services that enable continued operation. Modifications may be proposed to resolve supportability problems, to reduce operational costs, or to extend the life of a system. These changes require SE assessment to avoid loss of system capabilities while under operation. The corresponding technical process is the Maintenance Process.

3.3.7 Retirement Stage

The Retirement Stage is where the system-of-interest and its related services are removed from operation. SE activities in this stage are primarily focused on ensuring that disposal requirements are satisfied. In fact, planning for disposal is part of the system definition during the Concept Stage. Experience in the 20th century repeatedly demonstrated the consequences when system retirement and disposal are not considered from the outset. Early in the 21st century, many countries have changed their laws to hold the creator of a system-of-interest accountable for proper end-of-life disposal of the system.

3.4 Life Cycle Approaches

3.4.1 Plan-Driven Methods

The requirements/design/build/test/deploy paradigm is considered the traditional way to build systems. On projects where it is necessary to coordinate large teams of people working in multiple companies, plan-driven approaches provide an underlying framework to provide discipline to the life cycle processes. Plan-driven methods are characterized by a systematic approach that adheres to specified processes as the system moves through a series of representations from requirements through design to finished product. Specific attention is given to the completeness of documentation, traceability from requirements, and verification of each representation after the fact.

The strengths of plan-driven methods are predictability, stability, repeatability, and high assurance. Process improvement focuses on increasing process capability through standardization, measurement, and control. These methods rely on the "master plans" to anchor their processes and provide project-wide communication. Historical data is usually carefully collected and maintained as inputs to future planning to make projections more accurate.[18]

Safety-critical products, such as the Therac-25 medical equipment described in Section 3.6.1, can only meet modern certification standards by following a thorough, documented set of plans and specification. Such standards mandate strict adherence to process and specified documentation to achieve safety or security. However, unprecedented projects or projects with a high rate of unforeseeable change, predictability, and stability often degrade, and a project may incur significant investment trying to keep documentation and plans up-to-date.

3.4.2 Incremental and Iterative Development

Incremental and iterative development (IID) methods have been in use since the 1960s.[19] They represent a practical and useful approach that allows a project to provide an initial capability followed by successive deliveries to reach the desired system-of-interest. The goal is to provide rapid value and responsiveness. This approach is generally presented in opposition to the perceived burden associated with using any process, including those defined in this handbook.

The IID approach is used when the requirements are unclear from the beginning or the customer wishes to hold the system-of-interest open to the possibilities of inserting new technology. Based on an initial set of assumptions, a candidate system-of-interest is developed and then assessed to determine if it meets user needs or requirements. If not, another evolutionary round is initiated, and the process is repeated until a system is delivered to satisfied stakeholders or until the organization decides to terminate the effort.

Most literature agrees that IID methods are best applied to smaller, less complex systems or to system elements. The focus is on flexibility and on allowing selected events to be taken out of sequence when the risk is acceptable. Tailoring in this way highlights the core activities of product development.

The features that distinguish IID from the plan-driven approaches are velocity and adaptability. While market strategies often emphasize that "time to market" or speed is critical, a more appropriate criterion is "velocity," which considers direction in addition to speed. By incorporating the customer into the working-level teams, the project receives continuous feedback that they are going in a direction that satisfies the customer's highest needs first. One downside is that reactive project management with a customer that often changes direction can result in an unstable, chaotic project. On one hand, this approach avoids the loss of large investments in faulty assumptions; on the other hand, emphasis on a tactical viewpoint may generate short-term or localized solution optimizations.

IIDs may also be "plan-driven" in nature when the requirements are known early in the life cycle but the development of the functionality is performed incrementally to allow for the latest technology insertion or potential changes in needs or requirements. A specific IID methodology called evolutionary development[20] is common in research and development (R&D) environments. Figure 3-8 illustrates how this approach was used in the evolution of the tiles for the NASA Space Shuttle.[10]

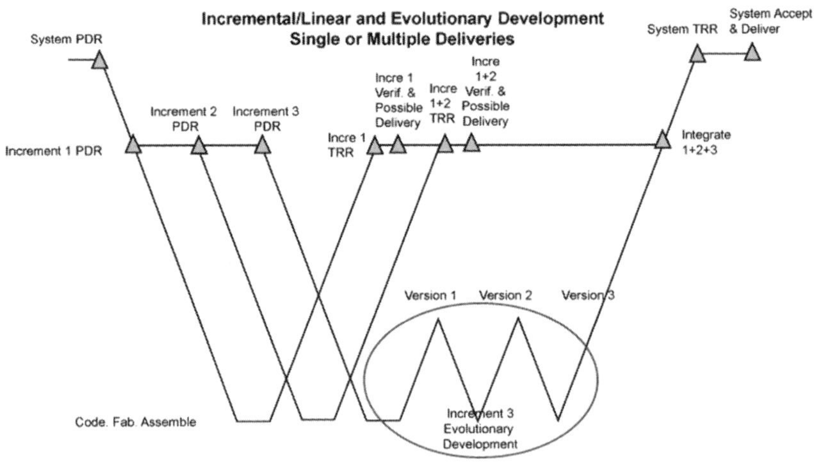

Figure 3-8 IID and Evolutionary Development[21]

3.4.3 Lean Development[21]

Systems engineering is regarded as an established, sound practice, but not always delivered effectively. Recent U.S. Government Accountability Office (GAO) and NASA studies of space systems[23,24,25,26] document major budget and schedule overruns. Similarly, recent studies by the MIT-based Lean Advancement Initiative (LAI)[27,28,29,30] have identified a significant amount of waste in government programs, reaching 70 percent of charged time. Most programs are burdened with some form of waste: poor coordination, unstable requirements, quality problems, and management frustration. This waste represents a vast productivity reserve in programs and major opportunities to improve program efficiency.

Lean Development and the broader methodology of Lean Thinking have their roots in the Toyota "Just-In-Time" philosophy, which aims at "producing quality products efficiently through the complete elimination of waste, inconsistencies, and unreasonable requirements on the production line."[31] Lean SE is the application of Lean Thinking to SE and related aspects of organization and project management. SE is focused on the discipline that enables flawless development of complex technical systems. Lean Thinking is a holistic paradigm that focuses on delivering maximum value to the customer and minimizing wasteful practices. Lean Thinking has been successfully applied in manufacturing, aircraft depots, administration, supply chain management, healthcare, and product development, including engineering.

Lean SE (LSE)[i] is the area of synergy between Lean Thinking and SE, with the goal to deliver the best life-cycle value for technically complex systems with minimal waste. Lean SE does not mean *less* SE. It means more and better SE with higher responsibility, authority, and accountability, leading to better, waste-free workflow with increased mission assurance. Under the Lean SE philosophy, mission assurance is non-negotiable, and any task that is legitimately required for success must be included, but it should be well-planned and executed with minimal waste.

Lean Thinking: "Lean thinking is the dynamic, knowledge-driven, and customer-focused process through which all people in a defined enterprise continuously eliminate waste with the goal of creating value."[32]

Lean Systems Engineering: The application of lean principles, practices and tools to SE to enhance the delivery of value to the system's stakeholders.

Three concepts are fundamental to the understanding of Lean Thinking: value, waste, and the process of creating value without waste (also known as Lean Principles).

3.4.3.1 Value

The value proposition in engineering programs is often a multi-year, complex, and expensive Acquisition Process involving thousands of stakeholders and resulting in hundreds or even thousands of requirements, which, notoriously, are rarely stable (even at the Request for Proposal [RFP] stage). In Lean SE, "value" is defined simply as mission assurance (i.e., the delivery of a flawless complex system, with flawless technical performance, during the product or mission development life cycle) and satisfying the customer and all other stakeholders, which implies completion with minimal waste, minimal cost, and the shortest possible schedule.

---

i The early use of the term LSE is sometimes met with concern that this might be a "re-packaged faster, better, cheaper" initiative, leading to cuts in SE at a time when the profession is struggling to increase the level and quality of SE effort in programs. The definitions provided herein should dispel those concerns and expose proper Lean Thinking.

"**Value** is a measure of worth (e.g., benefit divided by cost) of a specific product or service by a customer, and potentially other stakeholders and is a function of (1) the product's usefulness in satisfying a customer need, (2) the relative importance of the need being satisfied, (3) the availability of the product relative to when it is needed, and (4) the cost of ownership to the customer."[28]

3.4.3.2 Waste in Product Development

The LAI classifies waste into seven categories: Over-Processing, Waiting, Unnecessary Movement, Over-Production, Transportation, Inventory, and Defects.[33]

Waste: "The work element that adds no value to the product or service in the eyes of the customer. Waste only adds cost and time."[34]

When applying Lean Thinking to SE and project planning, consider each waste category and identify areas of wasteful practice. The following illustrates some waste considerations for SE practice in each of the LAI waste classifications.[32]

1. *Over-Processing* – Processing more than necessary to produce the desired output. Consider how projects "over do it" and expend more time and energy than needed:
    - Too many hands on the “stuff” (material or information)
    - Unnecessary serial production
    - Excessive/ custom formatting or reformatting
    - Excessive refinement, beyond what is needed for Value.
2. *Waiting* – Waiting for material or information, or information or material waiting to be processed. Consider "things" that projects might be waiting for to complete a task:
    - Late delivery of material or information
    - Delivery too early – leading to eventual rework.
3. *Unnecessary Movement* – Moving people (or people moving) to access or process material or information. Consider any unnecessary motion in the conduct of the task:
    - Lack of direct access – time spent finding what you need
    - Manual intervention.

4. *Over-Production* – Creating too much material or information. Consider how more "stuff" (i.e., material or information) is created than needed:
    - Creating unnecessary data and information
    - Information over-dissemination; pushing data.
5. *Transportation* – Moving material or information. Consider how projects move "stuff" from place to place:
    - Unnecessary hand-offs between people
    - Shipping "stuff" (pushing) when not needed
    - Incompatible communication – lost transportation through communication failures.
6. *Inventory* – Maintaining more material or information than is needed. Consider how projects stockpile information or materials:
    - Too much "stuff" built-up
    - Complicated retrieval of needed "stuff"
    - Outdated, obsolete information.
7. *Defects* – Errors or mistakes causing the effort to be redone to correct the problem. Consider how projects go back and do it again:
    - Lack of adequate review, verification, or validation
    - Wrong or poor information.

3.4.3.3 Lean Principles

Womack[34] captured the process of creating value without waste into six Lean principles.[ii] The principles are abbreviated as Value, Value Stream, Flow, Pull, Perfection, and Respect-for-People (see Figure 3-9), and are defined in detail below.

When applying Lean Thinking to SE and project planning, evaluate project plans, engineering processes, and organization behaviors using the Lean Principles. Consider how the customer defines *value* in the products and processes, then describe the *value stream* for creating products and processes, optimize *flow* through that value stream and eliminate waste, encourage *pull* from that value stream, and strive to *perfect* the value stream to maximize value to the customer. These activities should all be conducted within a foundation of *respect* for customers, stakeholders, and project team members.

[ii] The original formulation had five principles; the sixth (the Respect-for-People Principle) was added at a later time.

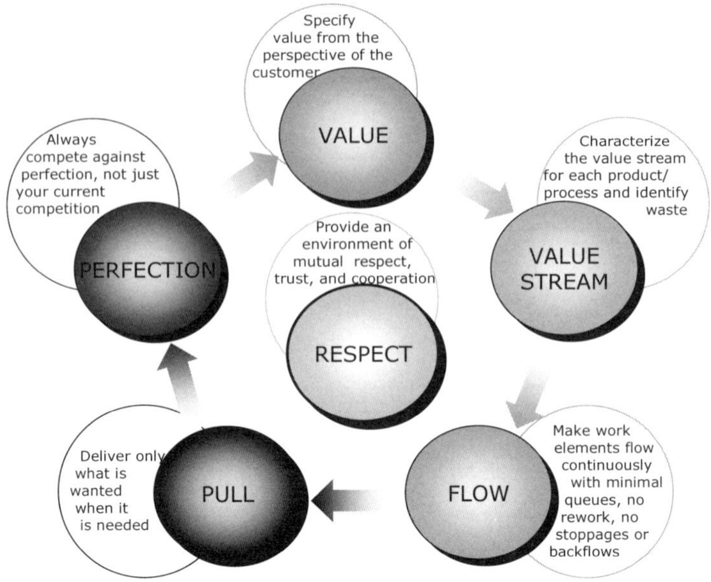

Figure 3-9 Lean Development Principles

In 2009, the INCOSE Lean SE Working Group released a new online product entitled *Lean Enablers for Systems Engineering* (LEfSE). It is a collection of 194 practices and recommendations formulated as "dos" and "don'ts" of SE based on Lean Thinking. The practices cover a large spectrum of SE and other relevant enterprise management practices, with a general focus on improving program value and stakeholder satisfaction and reducing waste, delays, cost overruns, and frustrations. LEfSE are grouped into the six Lean Principles. The LEfSE are not intended to become a mandatory practice; instead, they should be used as a checklist of good practices. The full text of the LEfSE is too long for the present text; therefore, only a brief summary is given herein. The full text is available online.[35]

- Under the *Value Principle*, enablers promote a robust process of establishing the value of the end-product or system to the customer with crystal clarity early in the program. The process should be customer-focused, involving the customer frequently and aligning employees accordingly.
- The enablers under the *Value Stream Principle* emphasize waste-preventing measures, solid preparation of the personnel and processes for subsequent efficient workflow and healthy relationships between stakeholders (e.g., customer, contractor, suppliers, and employees); detailed program planning; frontloading; and use of

leading indicators and quality measures. Systems engineers should prepare for and plan all end-to-end linked actions and processes necessary to realize streamlined value, after eliminating waste.

- The *Flow Principle* lists enablers that promote the uninterrupted flow of robust quality work and first-time right products and processes; steady competence instead of hero behavior in crises; excellent communication and coordination; concurrency; frequent clarification of the requirements; and making program progress visible to all.
- The enablers listed under the *Pull Principle* are a powerful guard against the waste of rework and overproduction. They promote pulling tasks and outputs based on customer need (including rejecting others as waste) and better coordination between the pairs of employees handling any transaction before their work begins so that the result can be first-time right.
- The *Perfection Principle* promotes excellence in the SE and organization processes; the use of the wealth of lessons learned from previous programs in the current program; the development of perfect collaboration policy across people and processes; and driving out waste through standardization and continuous improvement. A category of these enablers calls for a more important role of systems engineers, with responsibility, accountability and authority for the overall technical success of the program.
- Finally, the *Respect-for-People Principle* contains enablers that promote the enterprise culture of trust, openness, honesty, respect, empowerment, cooperation, teamwork, synergy, and good communication and coordination, and enable people for excellence.

The LEfSE were developed by 14 experienced INCOSE practitioners organized into two teams, including recognized industry leaders in Lean and System Engineering, academia, and government (from the United States, United Kingdom, and Israel), with cooperation from the 100-member strong international INCOSE Lean SE Working Group. The product has been evaluated by surveys and by comparisons with recent programmatic recommendations by GAO and NASA.[23,24,25,26] The surveys and comparisons indicate that the LEfSE, while not widely used by industry, are regarded as important for program success and are consistent with the GAO and NASA recommendations, though significantly more detailed and comprehensive.

The LEfSE are not intended to become a mandatory tool. Instead, they should be used as a checklist of good, holistic practices. Some enablers are intended for top enterprise managers, some for programs, and others for line employees. Some are more actionable than others, and some are easier to implement than others. Some enablers may require changes in company policies and culture. However, employee awareness of even those least actionable and most difficult to implement enablers should improve the thinking at work.

A formal online process of continuous improvement and periodic new releases of the LEfSE has been established and will be implemented as new knowledge and experience becomes available. A comprehensive article on the history of Lean SE, the development process of LEfSE, the full text of the enablers, the surveys, and industrial examples can be found in "Lean Enablers for Systems Engineering," accepted for publication by the *Journal of Systems Engineering* in 2009.[35]

3.4.4 Agile Development[36]

The preceding discussions have emphasized the benefits of orderly, hierarchical baseline progression followed by a corresponding verification sequence. Recognizing that the development process may require more flexibility in some circumstances, the Agile Development approach provides a tailoring framework, based on opportunity to simplify control methods and to assess the risks in so doing. The extent of tailoring is determined by whether the opportunity to shorten the project cycle is worth the risk of doing development steps out of sequence or in parallel.

The Agile Alliance (www.agilealliance.com) is dedicated to: developing iterative and agile methods, seeking a faster and better approach to software and system development, and challenging more traditional models. There are many articles describing the agile concepts.[37] The key objective is flexibility, and allowing selected events to be taken out of sequence (see Figure 3-10 and Figure 3-11) when the risk is acceptable.

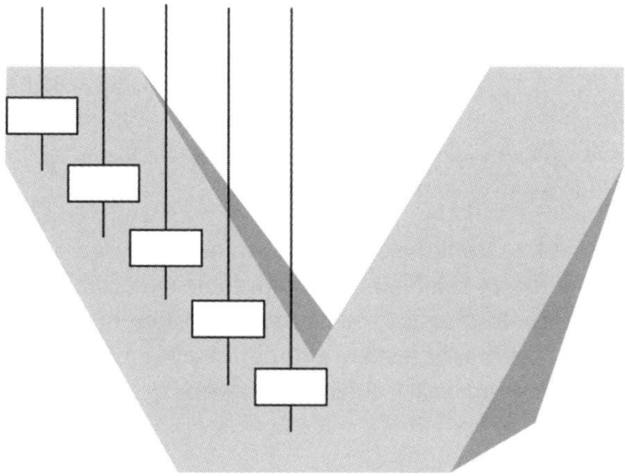

Figure 3-10 Hierarchical Baseline Elaboration

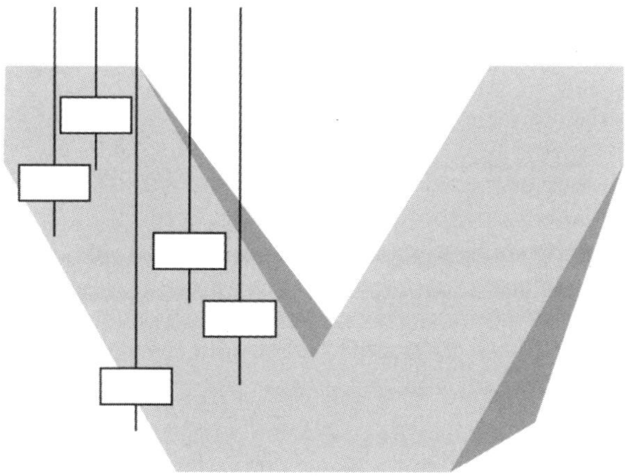

Figure 3-11 Non-hierarchical Baseline Elaboration

Similar to IID, Agile Development also distinguishes itself from conventional approaches through velocity and adaptability. Agile Development is in-process validation in action. By adopting the following seven key practices of agile SE (expanded upon in Craig Larman's, *Applying UML and Patterns: An Introduction to Object-Oriented Analysis and Design and Iterative Development*)[38], any organization can improve its velocity to customer satisfaction.

1. The project team understands, respects, works, and behaves within a defined SE process. The process is systemic in the organization and implicit to the participants.

2. The project is executed as fast as possible with minimum down time or staff diversion during the project. Every opportunity is exercised to move the project forward, especially for the critical path activities.

3. All key players are physically or electronically collocated. Other contributors are available on-line 24x7.

4. There is a strong bias for automatically generated electronic documentation. Engineers rely on their tools and their "Electronic Engineering Notebooks" to record decision rationale. Artifacts for operations and replication are done only if necessary—not to support an existing bureaucracy or policy. Notebooks are team property and are available to all.

5. Baseline management and change control is achieved by formal, oral agreement based on "make a promise, keep a promise" discipline—participants hold each other accountable. Decision gate agreements are confirmed with a binding handshake. Formality relates to the binding of the action not the amount of documentation.

6. Opportunity exploration and risk reduction are accomplished by expert consultation and rapid model verification, coupled with close customer collaboration. Software development is done in a rapid development environment, while hardware is developed in a multi-disciplined model shop. There is no resistance or inertia to securing expert help; it is sought rather than resisted.

7. A culture of constructive confrontation pervades the project organization. Issues are actively sought. Anyone can identify an issue and pass it on to the most likely solver. No issue is left unresolved. The team takes ownership for success; it is never "someone else's responsibility."

Agile Development principles (adapted for SE) are as follows:[39]

1. The highest priority is to satisfy the customer through early and continuous delivery of valuable software [and other system elements].

2. Welcome changing requirements, even late in development. Agile processes harness change for the customer's competitive advantage.

3. Deliver working software [and other system elements] frequently, from a couple of weeks to a couple of months, with a preference to the shorter timescale.

4. Business people and developers must work together daily throughout the project.
5. Build projects around motivated individuals. Give them the environment and support they need, and trust them to get the job done.
6. The most efficient and effective method of conveying information to and within a development team is face-to-face conversation.
7. Working software [and other system elements] is the primary measure of progress.
8. Agile processes promote sustainable development. The sponsors, developers, and users should be able to maintain a constant pace indefinitely.
9. Continuous attention to technical excellence and good design enhances agility.
10. Simplicity--the art of maximizing the amount of work not done--is essential.
11. The best architectures, requirements, and designs emerge from self-organizing teams.
12. At regular intervals, the team reflects on how to become more effective, then tunes and adjusts its behavior accordingly.

3.5 What is Best for Your Organization?

Conway's law[40] suggests that effective systems design emerges from system-oriented organizations and the processes employed therein. One of the earliest books on SE management[41] identified three simple criteria for such an organization: facilitate communications, streamline controls, and simplify paperwork. The way to effective SE management is not "in the direction of formal, formidable, massive documentation. It does, however, reside in the direction of creating a total environment which is conducive to the emergence and effective utilization of creative and inventive talents oriented toward achieving a system approach with a minimum of management encumbrances."[42]

Whenever someone (be it an individual or a company) wants to reach a desired end, they must perform a series of actions or operations. Further, they must consider the order of those actions, their dependencies, who will perform them, what they require and what they will generate, how long it will take to complete them, and what tools they will employ. Thus, individuals and organizations follow processes, be they predefined or ad hoc. Because process components (activities, products, agents, tools) and their interactions (information flow, artifacts flow, control, communication, timing, dependencies, and concurrency) can vary, processes will differ—even if the performing organizations have the same level, scope, and goal.

So why should an organization care about processes?[43] In short, to better understand, evaluate, control, learn, communicate, improve, predict, and certify the work performed. For a given organizational level, the processes vary with the project's goals and available resources. At a high level, the company's business strategy determines the business approach, with the main goals of time to market, minimum cost, or higher quality and customer satisfaction setting the priorities. Similarly, the company's size; the number, knowledge, and experience of people[44] (both engineers and support personnel); and hardware resources determine how to achieve those goals. The application domain and the corresponding system requirements, together with other constraints, form another important factor in defining and applying processes.

So what really is best for my organization? The answer is that it depends on the situation. Depending on the perspective, different processes are defined for entire organizations, teams, or individuals. A "one size fits all" approach does not work when defining processes, thus organizations must continuously document, define, measure, analyze, assess, compare, and change processes to best meet project goals. One would hardly expect to find the same processes used in a startup e-commerce company as in NASA. The intended goal shapes a process in terms of scope (namely, the stages and activities covered) and organizational level. In any case, the selected processes should help guide people on what to do—how to divide and coordinate the work—and ensure effective communication. Coordination and communication, for example, form the main problems in large projects involving many people, especially in distributed projects where people cannot communicate face to face.[45]

3.6 Introduction to Three Case Studies

Real-world examples that draw from diverse industries and types of systems are provided throughout this handbook. Three case studies have been selected to illustrate the diversity of systems to which SE principles and practices can be applied; medical therapy equipment, a bridge, and a super-high-speed train. They represent examples of failed, successful, and prototype systems that all define(d) the state-of-the-art. These studies may be categorized as medical, infrastructure, and transportation applications; in the manufacturing and construction industry domains; with and without software elements; complex; and subject to scrutiny both in the development and Utilization Stages as all have a need to be safe for humans and are constrained by government regulations.

3.6.1 Case 1: Radiation Therapy; the Therac-25

Therac-25, a dual-mode medical linear accelerator (LINAC), was developed by the medical division of Atomic Energy Commission Limited (AECL) of Canada, starting in 1976. The completely computerized system became commercially available in 1982. This new machine could be built at lower production cost, resulting in lower prices for the customers. However, a series of tragic accidents led to the recommended recall, and discontinuation of production of the system.

The Therac-25 was a medical LINAC, or particle accelerator, capable of increasing the energy of electrically charged atomic particles. LINACs accelerate charged particles by introducing an electric field to produce particle beams (i.e., radiation), which are then focused by magnets. Medical LINACs are used to treat cancer patients by exposing malignant cells to radiation. Since malignant tissues are more sensitive than normal tissues to radiation exposure, a treatment plan can be developed that permits the absorption of an amount of radiation that is fatal to tumors but causes relatively minor damage to surrounding tissue.

Therac-25 was a revolutionary design compared to its predecessors, Therac-6 and Therac-20, both with exceptional safety records. It was based on a double-pass concept that allowed a more powerful accelerator to be built into a compact and versatile machine. AECL designed Therac-25 to fully utilize the potential of software control. While Therac-6 and Therac-20 were built as stand-alone machines and could be operated without a computer, Therac-25 depended on a tight integration of software and hardware. In the new, tightly coupled system, AECL used software to monitor the state of the machine and to ensure its proper operations and safety. Previous versions had included independent circuits to monitor the status of the beam as well as hardware interlocks that prevented the machine from delivering radiation doses that were too high, or from performing any unsafe operation that could potentially harm the patient. In Therac-25, AECL decided not to duplicate these hardware interlocks since software already performed status checks and handled all the malfunctions. This meant that the Therac-25 software had far more responsibility for safety than the software in the previous models. If in the course of treatment, the software detected a minor malfunction it would pause the treatment. In this case, the procedure could be restarted by pressing a single "proceed" key. Only if a serious malfunction was detected was it required to completely reset the treatment parameters to restart the machine.

Software for Therac-25 was developed from the Therac-20's software, which was developed from the Therac-6's software. One programmer, over several years, evolved the Therac-6 software into the Therac-25 software. A stand-alone, real-time operating system was added along with application software written in assembly language and tested as a part of the Therac-25 system operation. In addition, significant adjustments had been made to simplify the operator interface and minimize data entry, since initial operators complained that it took too long to enter a treatment plan.

At the time of its introduction to market in 1982, Therac-25 was classified as a Class II medical device. Since the Therac-25 software was based on software used in the earlier Therac-20 and Therac-6 models, Therac-25 was approved by the Federal Drug Administration under Pre-Market Equivalency.

Six accidents involving enormous radiation overdoses to patients took place between 1985 and 1987. Tragically, three of these accidents were the direct cause of the death of the patient. This case is ranked in the top ten worst software-related incidents on many lists. Details of the accidents and analysis of the case is available from many sources.[46,47,48]

3.6.2 Case 2: Joining Two Countries; the Øresund Bridge

The Øresund Region is composed of eastern Denmark and southern Sweden and since 2000 is linked by the Øresund Bridge. The area includes the two major cities Copenhagen and Malmö, has a population of 3 million, and counts as Europe's eighth largest economic center. One fifth of the total Danish and Swedish Gross National Product (GNP) is produced in the region. The official name of the bridge is translated "the Øresund Connection" to underscore the full integration of the region. For the first time ever, Sweden is joined permanently to the mainland of Europe by a 10-minute drive or train ride. The cost for the entire Øresund Connection construction was calculated at 30.1 billion DKK (3 billion USD), and the investment is expected to be paid back by 2035.

The Øresund Bridge is the world's largest composite structure, has the longest cable-stayed bridge span in the world carrying motorway and railway traffic, and boasts the highest freestanding pylons. The 7.9 km (5 mi) long bridge crosses the international navigation route between the Baltic Sea and the North Sea. A cable-stayed high bridge rises 57 m (160 feet) above the surface of the sea, with a main span of 490 m (0.3 miles). Both the main span and the approach bridges are constructed as a two-level composite steel-concrete structure. The upper deck carries a four-lane motorway, and the lower deck carries a two-track railway for both passenger trains and freight trains. The rest of the distance is spanned by the artificial island Peberholm ("Pepper" islet, named to complement the Saltholm islet to the north) and a tunnel on the

Danish side that is the longest immersed concrete tunnel in the world. Since completion, Peberholm has become a natural habitat for colonies of rare birds, one of the largest of its kind in Denmark and Sweden.

Nations other than Denmark and Sweden also contributed to this project. Canada provided a floating crane, aptly named Svanen (the swan), to carry prefabricated bridge sections out to the site and place them into position. Forty-nine steel girders for the approach bridges were fabricated in Cádiz, Spain. A specially designed catamaran was built to handle transportation of the foundations for the pylons, which weighed 19,000 tons each.

The project began with well-defined time, budget, and quality constraints. The design evolved over more than seven years, from start to delivery of final documentation and maintenance manuals. More than 4000 drawings were produced. The consortium dealt with changes, as necessary, using a combination of technical competence and stakeholder cooperation. Notably, there were no disputes and no significant claims against the owners at the conclusion, and this has been attributed to the spirit of partnership.

From the beginning, the owners defined comprehensive requirements and provided definition drawings as part of the contract documents to ensure a project result that not only fulfilled the quality requirements on materials and workmanship, but also had the envisioned appearance. The contractor was responsible for the detailed design and for delivering a quality-assured product in accordance with the owners' requirements. The following are representative of the requirements levied at the start of the project:

- Schedule: Design life 100 years; Construction time Apr. 1996 - Apr. 2000
- Railway: Rail load UIC 71; Train speed 200 km/h
- Motorway: Road axle load 260 kN; Vehicle speed 120 km/h
- Ambient environment: Wind speed (10 min) 61 m/s; Wave height 2.5 m; Ice thickness 0.6 m; Temperature +/- 27°C
- Ship impact: to pylons 560 MN; to girder 35 MN.

In addition to established requirements, this project crossed national boundaries and was thereby subject to the legislations of each country. Technical requirements were based on the Eurocodes, with project specific amendments made to suit the national standards of both countries. Special safety regulations were set up for the working conditions, meeting the individual safety standards of Denmark and Sweden.

The railway link introduced yet another challenge. In Denmark, the rail traffic is right-handed, as on roadways, whereas the trains in Sweden pass on the left-hand side. The connection needed to ensure a logical transition between the two systems, including safety aspects. In addition, the railway power supply differs between the two countries, thus it was necessary to develop a system that could accommodate power supply for both railway systems and switch between them on the fly.

The design of a major cable-stayed bridge with approach spans for both road and railway traffic involves several disciplines, including but not limited to: geotechnical engineering, aerodynamics, foundation engineering, wind tunnel tests, design of piers and pylons, design of composite girders, design of cables and anchorages, design of structural monitoring system, ship impact analysis, earthquake analysis, analysis of shrinkage and creep of concrete, ice loads analysis, fatigue analysis, pavement design, mechanical systems, electrical systems, comfort analyses for railway passengers, traffic forecast, operation and maintenance aspects, analysis of construction stages, risk analysis for construction and operation, quality management, and environmental studies and monitoring.

Comprehensive risk analyses were carried out in connection with the initial planning studies, including specification of requirements to secure all safety aspects. Important examples of the results of these studies for the Øresund Bridge were as follows:

- Navigation span was increased from 330 m to 490 m
- The navigation channel was realigned and deepened to reduce ship groundings
- Pier protection islands were introduced to mitigate bridge/ship accidents.

Risks were considered in a systematic way, using contemporary risk analysis methods such as functional safety analyses using fault tree and "what if" techniques. Three main issues were considered under the design-build contract:

- General identification and assessment of construction risks
- Ship collision in connection with realignment of navigation channel
- Risks in connection with 5 years bridge operation by contractor.

A fully quantified risk assessment of the human safety and traffic delay risks was carried out for a comprehensive list of hazards, including: fire, explosion, train collisions and derailments, road accidents, ship collisions and groundings, aircraft collisions, environmental loads beyond design basis, and toxic spillages. An example of a consequence of this analysis was the provision of passive fire protection on the tunnel walls and ceilings.

Both Denmark and Sweden are proud of being among the cleanest industrial countries in the world. Their citizens, and therefore the politicians, would not allow for any adverse environmental impact from the construction or operation of a bridge. The Great Belt and Øresund Strait both constitute corridors between the salty Kattegat and the sweeter water of the Baltic Sea. Any reduction in water exchange would reduce the salt content and, therefore, the oxygen content of the Baltic Sea and would alter its ecological balance. The Danish and Swedish Authorities decided that the bridge should be designed in such a way that the flow-through of water, salt, and oxygen into the Baltic was not affected. This requirement was designated the zero solution. To limit impacts on the local flora and fauna in Øresund during the construction, the Danish and Swedish authorities imposed a restriction that the spillage of seabed material from dredging operations should not exceed 5% of the dredged amounts. The zero solution was obtained by modeling with two different and independent hydrographical models.

In total, 18 million cubic meters of seabed materials were dredged. All dredged materials were reused for reclamation of the artificial peninsula at Kastrup and the artificial island, Peberholm. A comprehensive and intensive monitoring of the environment was performed to ensure and document the fulfillment of all environmental requirements. In their final status report from 2001 the Danish and Swedish Authorities concluded that the zero solution as well as all environmental requirements related to the construction of the link had been fulfilled. Continual monitoring of eel grass and common mussels showed that, after a general but minor decline, populations had recovered by the time the bridge was opened. Overall, the environment paid a low price at both Øresund and the Great Belt because it was given consideration throughout the planning and construction stages of the bridges.

This award-winning bridge is the subject of numerous articles and a PhD thesis, where details of the construction history and collaboration among all the stakeholders are provided.[49,50,51] A National Geographic video documentary on the planning and construction of the Øresund Bridge is available on-line at http://www.great-engineering.com/oresund-bridge-megastructures.

3.6.3 Case 3: Prototype System; The Super-High-Speed Train in China

Shanghai Transrapid is the first commercial high-speed commuting system using the state-of-the-art electromagnetic levitation (or maglev) technology. The train runs from Shanghai's financial district to Pudong International

Airport, and the total track length is about 30 kilometers (20 miles). The train takes 7 minutes and 20 seconds to complete the journey, can reach almost 320 km/h (200 mph) in 2 minutes, and reaches its maximum speed of 430 km/h (267 mph) within 4 minutes. The Shanghai Transrapid project cost 10 billion Yuan (1.2 billion USD) and took 2.5 years to complete. Construction began in March 2001, and public service commenced on January 1, 2003. Critics argue that the speed over such a short distance is unnecessary and that the line may never recoup this cost.

Prior to this installation, many countries had argued over the feasibility of maglev trains. They do not have wheels or use a traditional rail. Rather, powerful magnets lift the entire train about 10 millimeters above the special track, called a guideway, which mainly directs the passage of the train. Electromagnetic force is used to make the train hover and to provide vertical and horizontal stabilization. The frequency, intensity, and direction of the electrical current in the track control the train's movement, while the power for the levitation system is supplied by the train's onboard batteries, which recharge whenever the train is moving. Maglev trains also do not have an onboard motor. The guideway contains a built-in electric motor that generates an electromagnetic field that pulls the train down the track. Putting the propulsion system in the guideway rather than onboard the trains, makes the cars lighter, which enables the train to accelerate quickly. The super-high speeds are attained largely due to the reduction of friction.

Despite the high speed, the maglev system runs more quietly than a typical commuter train, consumes less energy, and is nearly impossible to derail because of the way the train's underside partially wraps around the guideway, like a giant set of arms hugging the train to the elevated platform. Passengers experience a comfortable and quiet ride due to the maglev technology and the specially designed window; noise level is less than 60 decibels at a speed of 300 km/h.

The Chinese authorities considered the economical operation, low energy consumption, less environmental impact, and high speed when choosing a solution suitable for ground transport between hubs that range from hundreds to over one thousand kilometers apart. But the same solution also needed to be suitable for modern mass rapid passenger transportation between a center city and adjacent cities. Despite the many advantages, in 1999 the technology was considered to be in an experimental stage—its technological superiority, safety and economic performance not yet proven by commercialized operation. The current line is the result of a compromise; it was built as a demonstration to verify the maturity, availability, economics, and safety of a high-speed maglev transportation system.

The basic technology to create a maglev system has been around since 1979, but until this project it had never been realized, mostly due to the expense of developing a new train system. Many experts believe that super-fast steel-wheel rail systems—such as those in France and Japan—have reached the limits of this technology and cannot go any faster. Maglev proponents describe the system as "the first fundamental innovation in the field of railway technology since the invention of the railway" and are watching proposals for maglev installations in Germany and the USA.[52,53,54,55]

3.7 References

1. ISO/IEC 15288:2008, p. 10
2. Ibid, p. 19
3. Used with permission of H. Stoewer
4. ISO/IEC 15288:2008, p. 10
5. Ibid, p. 19
6. Adapted from ISO/IEC 15288:2002, page 57, Table D-1
7. Contributed by Hironori Washizaki, National Institute of Informatics, Waseda University, Japan
8. Adapted from figure provided by Kevin Forsberg, Center for Systems Management
9. Adapted from Forsberg, K., H. Mooz, H. Cotterman, *Visualizing Project Management*, 3rd Ed., J. Wiley & Sons, 2005
10. Ibid p. 111
11. Forsberg, K., "'If I Could Do That, Then I Could…' System Engineering in a Research and Development Environment," *Proceedings of Fifth NCOSE Summer Symposium*, 1995
12. *Pre-Milestone A and Early-Phase Systems Engineering*, National Research Council of the National Academies (USA), Washington, D.C.: The National Academies Press, 2008, <http://www.nap.edu>
13. "NASA's Ambitious New Mars Rover Is Too Costly, Critics Say," *Washington Post*, February 11, 2009
14. Used with permission from United Media
15. SysTest, *Systems Verification, Validation and Testing Methodology Guidelines*, Contract: G1RD-CT-2002-00683, <http://www.incose.org/secoe/0105.htm>
16. Forsberg, K., H. Mooz, H. Cotterman, *Visualizing Project Management*, 3rd Ed., J. Wiley & Sons, 2005, p. 115
17. ISO/IEC 15288:2008, p. 11
18. Boehm, B., and R. Turner, *Balancing Agility and Discipline*, Addison-Wesley, 2004
19. Larman, C. V. Basili, "Iterative and Incremental Development: A Brief History," *IEEE Software*, May/June 2003
20. Gilb, T., *Competitive Engineering,* Elsevier, 2005, p. 358
21. This section adapted from Oppenheim, Bohdan W., "Lean Enablers for Systems Engineering," *CrossTalk: The Journal of Defense Software Engineering*, July/August 2009, <http://www.stsc.hill.af.mil>
22. Forsberg, K., H. Mooz, H. Cotterman, *Visualizing Project Management*, 3rd Ed., J. Wiley & Sons, 2005, p. 358
23. GAO, "Defense Acquisitions – Assessments of Selected Weapon Programs," GAO-07-4065SP. Washington, D.C., March 2008 < http://www.gao.gov/new.items/d08467sp.pdf>
24. GAO, "Best Practices – Increased Focus on Requirements and Oversight Needed to Improve DOD's Acquisition Environment and Weapon System Quality," GAO-08-294, Washington, D.C., February 2008 < http://www.gao.gov/new.items/d08294.pdf>

25. GAO, "Space Acquisitions – Major Space Programs Still at Risk for Cost and Schedule Increases," GAO-08-552T, Washington, D.C., 4 March 2008 <http://www.gao.gov/new.items/d08552t.pdf>
26. "NASA Pilot Benchmarking Initiative: Exploring Design Excellence Leading to Improved Safety and Reliability," Final Report, October 2007
27. LAI, "Phase I," 1 Jan, 2009 <http://lean.mit.edu/index.php?/about-lai/history/phase-one.html>
28. McManus, Hugh L., "Product Development Value Stream Mapping Manual," LAI Release Beta, MIT, April 2004
29. Slack, Robert A., "Application of Lean Principles to the Military Aerospace Product Development Process," *Master of Science – Engineering and Management Thesis*, MIT, December 1998
30. Oppenheim, Bohdan W, "Lean Product Development Flow," Journal of Systems Engineering, Vol. 7, No. 4, 2004
31. "Just-In-Time – Productivity Improvement," *Toyota Vision and Philosophy*, 2009, <http://www2,toyota,co,jp/en/vision/production_system>
32. Murman, E. M., et al, *Lean Enterprise Value*, Palgrave, 2002
33. McManus, Hugh L, "Product Development Transition to Lean (PDTTL) Roadmap," LAI Release Beta, MIT, March 2005
34. Womack, James P., and Daniel T, Jones, *Lean Thinking*, New York: Simon & Schuster, 1996
35. Oppenheim, Bohdan W., "Lean Enablers for Systems Engineering," *CrossTalk Defense Journal*, July/August 2009; *Systems Engineering: Journal of the International Council on Systems Engineering*, accepted 2009
36. Forsberg, K., H. Mooz, H. Cotterman, *Visualizing Project Management*, 3rd Ed., J. Wiley & Sons, 2005
37. see <http://www.agilealliance.com/articles>
38. Craig Larman, *Applying UML and Patterns: An Introduction to Object-Oriented Analysis and Design and Iterative Development*, 3rd ed., Upper Saddle River, NJ: Prentice Hall, 2005
39. "Principles behind the Agile Manifesto," 15 December 2009, <http://www.agilemanifesto.org/principles.html>
40. Conway, Melvin E, "How do Committees Invent?", *Datamation* 14 (5): 28–31, April, 1968, <.http://www.melconway.com/research/committees.html>
41. Chase, W, P., *Management of System Engineering,* John Wiley & Sons, 1974
42. Ibid, p 147
43. McConnell, Steve, "The Power of Process," *IEEE Computer*, May 1998
44. Cockburn, A., "Selecting a Project's Methodology," *IEEE Software*, July/Aug 2000, p. 64-71
45. Lindvall, Mikael, and Ioana Rus, "Process Diversity in Software Development," *IEEE Software* July/Aug 2000
46. Jacky, Johathan, "Programmed for Disaster," *The Sciences* 29 (1989): 22-27
47. Leveson, N., and C. S. Turner, "An Investigation of the Therac-25 Accidents," *IEEE Computer*, Vol. 26, No. 7, 1993, pp. 18-41
48. Porrello, Anne Marie, "Death and Denial: The Failure of the THERAC-25, A Medical Linear Accelerator," <http://www.csc.calpoly.edu/~jdalbey/SWE/Papers/THERAC25.html>
49. Nissen, J., "The Øresund Link," *The Arup Journal*, 31(2): February 2996, pp. 37-41
50. Jensen, J. S., "The Øresund Bridge – Linking Two Nations," <http://www.cowi.dk>
51. Website of the Øresund Bridge Consortium, <http://osb.oeresundsbron.dk>
52. "China's Supertrain Takes to Tracks," *BBC News Online: Asia-Pacific*, 9 August 2002, 29 May 2003, <http://news.bbc.co.uk/2/hi/asia-pacific/2182975.stm>
53. McGrath, Dermot, "China Awaits High Speed 'Maglev,'" *Wired News*, 23 January 2003, <http://www.wired.com/news/technology/0.1282.57163.00.html>
54. "Transrapid," Transrapid International, 23 May 2003 <http://www.transrapid.de>
55. Website of Shanghai Maglev Transportation Development Company, <http://www.smtdc.com/en/gycf4.asp>

4 Technical Processes

The ISO/IEC 15288:2008 Technical Processes are invoked throughout the life-cycle stages of a system. Technical Processes are defined in ISO/IEC 15288:2008 as follows:

> *The Technical Processes are used to define the requirements for a system, to transform the requirements into an effective product, to permit consistent reproduction of the product where necessary, to use the product to provide the required services, to sustain the provision of those services and to dispose of the product when it is retired from service.*
>
> *The Technical Processes define the activities that enable organization and project functions to optimize the benefits and reduce the risks that arise from technical decisions and actions. These activities enable products and services to possess the timeliness and availability, the cost effectiveness, and the functionality, reliability, maintainability, producibility, usability and other qualities required by acquiring and supplying organizations. They also enable products and services to conform to the expectations or legislated requirements of society, including health, safety, security and environmental factors.*[1]

Figure 1-1 illustrates the relationship of the Technical Processes to the Project, Agreement, and Organizational Project-Enabling Processes. Without the Technical Processes, the risk of project failure would be unacceptably high.

In his opening keynote at the *15th Annual INCOSE International Symposium*, Riley Duren of Jet Propulsion Laboratory, California, stressed that SE is a way of thinking about and solving challenges and that systems engineers are the GLUE that hold the elements of complex space programs together[2]. To achieve good results, systems engineers involve themselves in nearly every aspect of a project, pay close attention to interfaces where two or more systems or system elements work together, and establish an interaction network with stakeholders and other organizational units of the organization. Figure 4-1 shows the critical interactions for systems engineers.

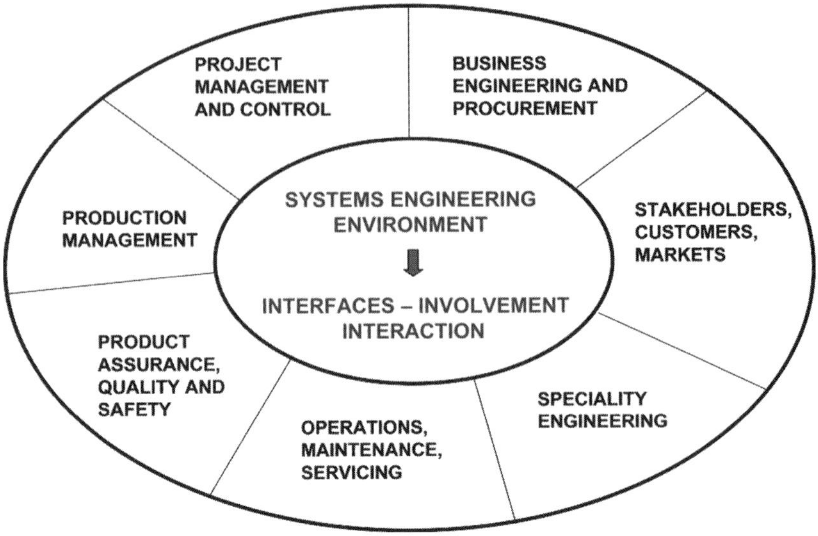

Figure 4-1 Key SE Interactions[3]

Technical Processes enable systems engineers to coordinate the interactions between engineering specialists, systems stakeholders and operators, and manufacturing. They also address conformance with the expectations and legislated requirements of society. These processes lead to the creation of a full set of requirements that address the desired capabilities within the bounds of performance, environment, external interfaces, and design constraints.

4.1 Stakeholder Requirements Definition Process

4.1.1 Overview

4.1.1.1 Purpose

As stated in ISO/IEC 15288:2008:

> *The purpose of the Stakeholder Requirements Definition Process is to define the requirements for a system that can provide the services needed by users and other stakeholders in a defined environment.*
>
> *It identifies stakeholders, or stakeholder classes, involved with the system throughout its life cycle, and their needs, expectations, and desires. It analyzes and transforms these into a common set of stakeholder requirements that express the intended interaction the system will have with its operational environment and that are the reference against which each resulting operational service is validated.*[4]

There is near unanimous agreement that successful projects depend on meeting the needs and requirements of the stakeholder/customer.

4.1.1.2 Description

A stakeholder is any entity (individual or organization) with a legitimate interest in the system. Typical stakeholders include users, operators, organization decision-makers, parties to the agreement, regulatory bodies, developing agencies, support organizations, and society-at-large. When direct contact is not possible, systems engineers find agents, such as marketing or non-governmental organizations, to represent the concerns of a class of stakeholders, such as consumers or future generations.

The stakeholder requirements govern the system's development and are an essential factor in further defining or clarifying the scope of the development project. If an organization is acquiring the system, this process provides the basis for the technical description of the deliverables in an agreement – typically in the form of a system-level specification and defined interfaces at the system boundaries. Figure 4-2 is the context diagram for this process.

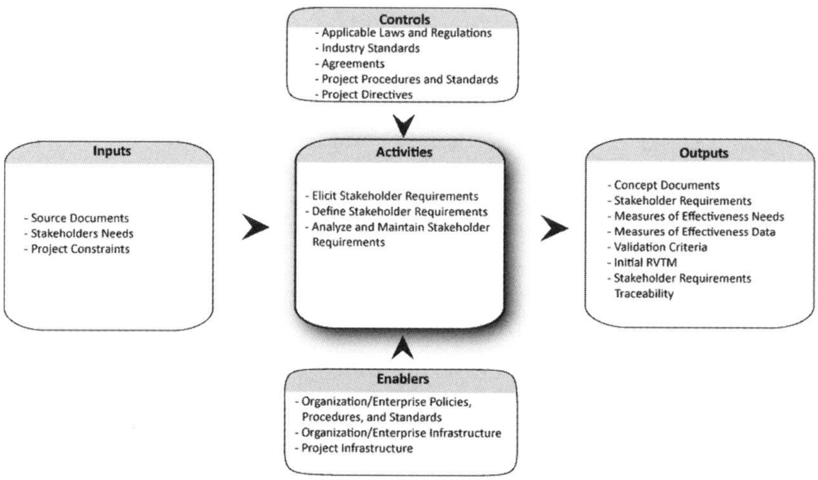

Figure 4-2 Context Diagram for Stakeholder Requirements Definition Process

4.1.1.3 Inputs

Inputs to the Stakeholder Requirements Definition Process include the following:

- *Source Documents* – Extract, clarify, and prioritize all of the written directives embodied in the source documents relevant to the particular stage of procurement activity.

- *Stakeholders' Needs* – Description of users' and other stakeholders' needs or services that the system will provide.
- *Project Constraints* – Includes cost, schedule, and solution constraints.

This process is governed by the following controls and enablers:

- *Applicable Laws and Regulations*
- *Industry Standards* – relevant industry specifications and standards
- *Agreements* – terms and conditions of the agreements
- *Project Procedures and Standards* – including project plans
- *Project Directives*
- *Organization/Enterprise Policies, Procedures, and Standards* – including guidelines and reporting mechanisms
- *Organization/Enterprise Infrastructure*
- *Project Infrastructure*.

4.1.1.4 Outputs

Outputs of the Stakeholder Requirements Definition Process establish the initial set of set of stakeholder requirements for project scope and associated agreements and include the following:

- *Concept Documents*
  - *Concept of Production* – Describes the way the system will be manufactured, including any hazardous materials used in the process.
  - *Concept of Deployment* – Describes the way the system will be delivered and installed.
  - *Concept of Operations (ConOps)* – Describes the way the system works from the operator's perspective. The ConOps includes the user description and summarizes the needs, goals, and characteristics of the system's user community. This includes operation, maintenance, and support personnel.
  - *Concept of Support* – Describes the desired support infrastructure and manpower considerations for maintaining the system after it is deployed. This includes specifying equipment, procedures, facilities, and operator training requirements.

  - *Concept of Disposal* – Describes the way the system will be removed from operation and retired, including the disposal of any hazardous materials used in or resulting from the process.

- *Stakeholder Requirements* – Formally documented and approved stakeholder requirements that will govern the project, including: required system capabilities, functions, and/or services; quality standards; and cost and schedule constraints.
- *Measures of Effectiveness Needs* – Measures of Effectiveness (MOEs) are the "operational" measures of success that are closely related to the achievement of the mission or operational objective being evaluated, in the intended operational environment under a specified set of conditions (i.e., how well the solution achieves the intended purpose).
- *MOE Data* – Data provided to measure the MOEs.
- *Validation Criteria* – May specify who will perform validation activities, and the environments of the system-of-interest.
- *Initial Requirements Verification and Traceability Matrix* (RVTM) – A list of requirements, their verification attributes, and their traces.
- *Stakeholder Requirements Traceability* – All stakeholder requirements should have bi-directional traceability, including to their source, such as the source document or the stakeholder need.

4.1.1.5 Process Activities

The Stakeholder Requirements Definition Process includes the following activities:

- *Elicit Stakeholder Requirements*
  - Identify stakeholders who will have an interest in the system throughout its entire life cycle.
  - Elicit requirements – what the system must accomplish and how well.
- *Define Stakeholder Requirements*
  - Define constraints imposed by agreements or interfaces with legacy enabling systems.
  - Build scenarios to define the concept documents; the range of anticipated uses of system products; the intended operational environment; and interfacing systems, platforms, or products. Scenarios help identify requirements that might otherwise be

  overlooked. Social and organizational influences also emerge from using scenarios.
  - Establish critical and desired system performance – thresholds and objectives for system performance parameters that are critical for system success and those that are desired but may be subject to compromise to meet the critical parameters.
  - Establish MOEs and suitability – measures that reflect overall customer/user satisfaction (e.g., performance, safety, reliability, availability, maintainability, and workload requirements).
- *Analyze and Maintain Stakeholder Requirements*
  - Analyze requirements for clarity, completeness, and consistency.
  - Negotiate modifications to resolve unrealizable or impractical requirements.
  - Validate, record, and maintain stakeholder requirements throughout the system life cycle and beyond for historical or archival purposes.
  - Establish and maintain a traceability matrix to document how the formal requirements are intended to meet the stakeholder objectives and achieve stakeholder agreement.

Common approaches and tips:

- Develop a description of the user community to provide common understanding across the effort and to validate the appropriateness of scenarios. A user description may cover the demographic group(s) to which a product will be marketed or the specific personnel categories that will be assigned to employ the system or otherwise benefit from its operation.
- Once stakeholders' requirements are established, formally place them under configuration control.
- Establish good relationships and open communications between systems engineers and stakeholders. This is helpful when negotiations begin to refine and clarify the set of requirements.
- Identify all stakeholders. It is critical to identify and include key system stakeholders in this process including the development/design team.
- Avoid designing a final solution or establishing unjustified constraints on the solution.

- Avoid accepting unrealistic or competing objectives.
- Write clearly and create statements with quantifiable values.[5]
- Capture source and rationale for each requirement.[6]

4.1.2 Elaboration

Within the context of ISO/IEC 15288:2008, requirements are specifically mentioned in two of the Technical Processes (i.e., Stakeholder Requirements Definition and Requirements Analysis) and are drivers for many of the system life-cycle processes. Depending on the system development model, stakeholder requirements capture should be conducted nominally once near the beginning of the development cycle or as a continuous activity. Regardless, the reason for eliciting and analyzing requirements is the same—understand the needs of the stakeholders well enough to support the Architectural Design Process.

4.1.2.1 Identify Users and Stakeholders

One of the biggest challenges in this activity is the identification of the set of stakeholders from whom requirements should be elicited. Customers and eventual end-users are relatively easy to identify, but regulatory agencies and other interested parties that may reap the consequences of the system-of-interest should also be sought out and heard. Stakeholders can include the interoperating systems and enabling systems themselves as these will usually impose constraints that need to be identified and considered. In sustainable development this includes finding representation for future generations. Figure 4-3 illustrates the range of potential stakeholders.

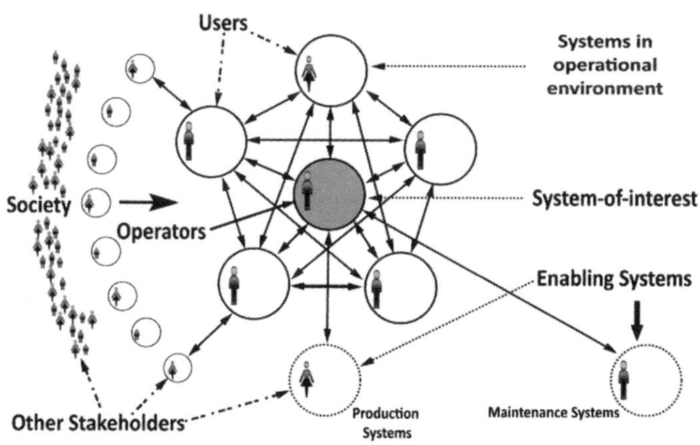

Figure 4-3 Requirements elicitation captures the needs of stakeholders across systems boundaries[7]

4.1.2.2 *Define Needs*

The initiation of a project begins with user need. Once a need is perceived and resources are committed to establish a project, it is possible to define the parameters of an acquisition and supply relationship (see Chapter 6).

Determining user need typically occurs in the Exploratory Research, Concept, and/or Development stage of the development life cycle since the "big picture" is developed and authorization to fund the development and select the developers/suppliers are accomplished. In many cases the integration of user views, which may not necessarily be harmonious, is accomplished by "committee action." However, this can lead to confusion and unsatisfactory decision making. As the SE process is applied, a common paradigm for examining and prioritizing available information and determining the value of added information can be created. Each of the user's views of the needed systems can be translated to a common top-level program and system description that is understood by all participants, and all decision making activities recorded for future examination. Under some circumstances, it may not be practical to elicit needs from the "user" but rather from the marketing organization or other surrogate.

Many tools and techniques can be used to elicit user requirements, such as marketing and technical questionnaires or surveys, focus groups, prototypes, and beta release of a product. Trade-off analysis and simulation tools can also be used to evaluate mission operational alternatives and select the desired mission alternative.

Systems engineering should support program and project management in defining what must be done and gathering the information, personnel, and analysis tools to define the mission or program objectives. This includes gathering customer inputs on "needs" and "wants," system/project constraints (e.g., costs, technology limitations, and applicable specifications/legal requirements), and system/project "drivers," such as capabilities of the competition, military threats, and critical environments. The set of recommended activities that follow are written for a complex project that meets a stated mission or goal; the word "product" can be substituted to apply these steps to commercial products:

1. Identify users and other stakeholders and understand their needs. Develop and document the new mission needs of all user organizations through user surveys.
2. Perform mission analysis to establish the operational environment, requirements functionality, and architecture and to assess existing capability.

3. Document the inadequacies or cost of existing systems to perform new mission needs.
4. If mission success is technology driven, develop concepts and document the new capabilities that are made possible by the introduction of new or upgraded technology. Document the tradeoffs in mission performance vs. technology steps.
5. Prepare a justification of the need for this mission compared to alternative missions competing for the same resources.
6. Prepare the necessary documents to request funding for the first program stage.
7. If system procurement is involved, develop the information needed to release an RFP, establish the selection criteria and perform a source selection.

The output of mission level activities should be sufficient definition of the operational need or ConOps to gain authorization and funding for program initiation. The output should also generate an RFP if the system is to be acquired through a contract Acquisition Process, or to gain authorization to develop and market the system if market driven. These outputs can be documented in a mission statement, a System Requirements Document (SRD), a Statement of Work (SOW), and/or an RFP. Contributing users rely on well-defined completion criteria to indicate the successful definition of user and stakeholder needs:

- User organizations have gained authorization for new system acquisition.
- Program development organizations have prepared a SOW, SRD, and gained approval for new system acquisition. If they are going to use support from outside the company, they have issued an RFP, and selected a contractor.
- Potential contractors have influenced the acquisition needs, submitted a proposal, and have been selected to develop and deliver the system.
- If the system is market driven, the marketing group has learned what consumers want to buy. For expensive items (e.g., aircraft) they have obtained orders for the new systems.
- If the system is market and technology driven, the development team has obtained approval to develop the new system from the corporation.

4.1.2.3 Capture Source Requirements

This section discusses methods for capturing requirements from user objectives and the customer's preliminary requirements. As shown in Figure 4-3, requirements come from multiple sources. As such, eliciting and capturing requirements constitutes a significant effort on the part of the systems engineer. The ConOps describes the intended operation of the system to be developed and helps the systems engineer understand the context within which requirements need to be captured and defined. Techniques for requirements elicitation include interviews, focus groups, the Delphi technique, and soft systems methodology. Tools for capturing and managing requirements are many and varied. The INCOSE Tools Database Working Group evaluates the relative merits of different products and maintains a database that is available from the INCOSE website.

The SE team leaders extract, clarify, and prioritize all of the written directives embodied in the source documents relevant to the particular stage of procurement activity. Examples of typical inputs include (but are not limited to):

1. New or updated customer needs, requirements, and objectives in terms of missions, ConOps, MOEs, technical performance, utilization environments, and constraints
2. Technology base data including identification of key technologies, performance, maturity, cost, and risks
3. Requirements from contractually cited documents for the system and its configuration items (CIs)
4. Technical objectives
5. Records of meetings and conversations with the customer.

The source requirements captured by carrying out this activity are only a portion of the total stakeholder requirements. As such, source requirements will be expanded by a number of activities designed to break down the broad requirements statements and reveal the need for additional clarification, which will lead to either revision of the written source material or additional source documents, such as meeting minutes. The ConOps definition activity, discussed in Section 4.1.2.5, may also reveal the need for additional clarification.

4.1.2.4 Initialize the Requirements Database

The primary objective of this activity is to establish a database of baseline system requirements traceable to the source needs and requirements to serve as a foundation for later refinement and/or revision by subsequent activities in the SE process. All parties involved in this activity (users, developing agencies, builders, support organizations, etc.) should maintain and contribute to this database. Prerequisites for the successful performance of this activity are:

1. Empower a systems analysis team with the authority and mission to carry out the activity.
2. Assign experienced Systems Engineer(s) to lead the team.
3. Assign experienced team members from relevant engineering, test, manufacturing, and operations (including logistics) disciplines to be available to the team.
4. Establish the formal decision mechanism (e.g., a design decision database) and any supporting tools; select and obtain necessary SE tools for the activity.
5. Complete the relevant training of team members in the use of tools selected for the activity.
6. Define the formats of the output deliverables from this activity (to permit the definition of any database schema tailoring that may be needed).

The requirements database must first be populated with the source documents that provide the basis for the total set of system requirements that will govern its design. The following guidance has proven helpful in establishing a Requirements Database:

1. Take the highest priority source document identified and ensure that it is recorded in the database in a manner such that each paragraph in the source document is recorded as a separate requirements object. Record information needed to trace each requirement back to the identity of:

    - The source document identity
    - The paragraph title
    - The sentence number.

    One reason for selecting a paragraph as the parent requirement is to better evaluate the impact of later changes since most changes to source documents are flagged by a change bar against paragraphs that have been modified or deleted.

2. Analyze the content of each parent requirement produced in the previous step. Based on its engineering content, determine the following:

- Does the parent requirement contain any information systems objectives? If it does, it should meet the following criteria:
    - Unambiguous
    - Non-conflicting with other requirements
    - Uniquely assignable to a single system function, architectural element, performance measurement index, or system constraint.

  If it meets the previous criteria, bypass the mini-steps below and move to the next parent requirement object. If the engineering content does not meet these criteria, determine a strategy for decomposing the parent requirement into separate but related pieces with the objective of meeting these criteria. This is accomplished as follows:

    1. Record information in the database to provide vertical traceability from the parent requirement object to the child requirement object using the Project Unique Identifier (PUID).
    2. Repeat the procedure with child requirements as necessary, creating and recording traceability to grandchild, great-grandchild, etc. down to the lowest level requirement. Stop fragmentation at the level when the decomposition objective has been achieved. This is called a leaf node requirement. Each branch of the tree may be decomposed down to a different level, depending on the complexity and the system acquisition approach (e.g., make or code everything or use suppliers or subcontractors for one or more branches). The process of flowing down requirements will eventually end at a configuration item (CI). A CI is a hardware, software, or composite item at any level in the system hierarchy designated for configuration management. CIs have four common characteristics:
        a. Defined functionality
        b. Replaceable as an entity
        c. Unique specification
        d. Formal control of form, fit, and function.

3. Repeat steps 1 and 2 for lower priority source documents.

Source documents used as inputs will include statements of user objectives, customer requirements documents, marketing surveys, systems analysis, concept analyses and others. These source or originating requirements should be entered in the database and disseminated to all team members assigned to the requirements analysis team. The information should also be accessible for rapid reference by other project personnel. As a minimum, this foundation must include the following:

1. Project requirements
2. Mission requirements
3. Customer specified constraints
4. Interface, environmental, and non-functional requirements
5. Unclear issues discovered in the Requirements Analysis Process
6. An audit trail of the resolution of the issues raised
7. V&V methods required by the customer
8. Traceability to source documentation
9. Substantiation (verification) that the database is a valid interpretation of user needs.

4.1.2.5 Establish the Concept of Operations

The word "scenario" is often used to describe a single thread of behavior; in other cases, it describes a superset of many single threads operating concurrently. Scenarios and what-if thinking are essential tools for planners who must cope with the uncertainty of the future. Scenario thinking can be traced back to the writings of early philosophers, such as Plato and Seneca.[8] As a strategic planning tool, scenario techniques have been employed by military strategists throughout history. Building scenarios serves as a methodology for planning and decision-making in complex and uncertain environments. The exercise makes people think in a creative way, observations emerge that reduce the chances of overlooking important factors, and the act of creating the scenarios enhances communications within and between organizations. Scenario building is an essentially human activity that may involve interviews with operators of current/similar systems, potential end users, and meetings of an Interface Working Group (IFWG). The results of this exercise can be captured in many graphical forms using modeling tools and simulations.

Creation or upgrade of a system shares the same uncertainty regarding future use and emergent properties of the system. The Stakeholder Requirements Definition Process suggests capturing the understanding of stakeholder needs in a series of concept documents (see Section 4.1.1.4), each focused on a specific life-cycle stage: Concept of Production, Concept of Deployment, ConOps, Concept of Support, and Concept of Disposal. A primary goal of a concept document is to capture, early in the system life cycle, an implementation-free understanding of stakeholders' needs by defining what is needed, without addressing how to satisfy the need. It captures behavioral characteristics required of the system in the context of other systems with which it interfaces, and captures the manner in which people will interact with the system for which the system must provide capabilities. Understanding these operational needs typically produces:

- A source of specific and derived requirements that meet the customer and user needs and objectives.
- Invaluable insight for Integrated Product Development Team (IPDT) members as they design, develop, verify, and validate the system.
- Diminished risk of latent system defects in the delivered operational systems.

If the system is for a military customer, there may be several required operational views of the system driven by architectural frameworks. These are defined, for example, in the U.S. Department of Defense Architecture Framework (DODAF) and in the UK Ministry of Defense Architecture Framework (MODAF).

The ConOps document, sometimes called the Operational Concept Document (OCD), defines the way the system will be used and must involve the input from a broad range of stakeholders, such as operations, maintenance, and management personnel. The document also defines any critical, top-level performance requirements or objectives (stated either qualitatively or quantitatively) and system rationale, and contains a preliminary functional block diagram of the system with only the top-level functional "threads" specified, as well as the roles and responsibilities and the set of skills needed for operations and maintenance of the system.

> Note: The U.S. DOD and others use the term ConOps to mean how the *enterprise* will operate. They use the term Operational Concept to discuss how the *system* will operate within the context of the enterprise. This handbook uses the term ConOps for both instances.

A ConOps document typically comprises the following:

- A top-level operational concept definition containing approved operational behavior models for each system operational mode (which can be documented as functional flow diagrams), supporting time lines, and event transcripts, which are fully traceable from source requirements

- Context diagrams
- Mission Analyses.

The primary objective is to communicate with the end user of the system during the early specification stages to ensure that operational needs are clearly understood and the rationale for performance requirements is incorporated into the decision mechanism for later inclusion in the system and lower level specifications. Interviews with operators of current/similar systems, potential users, IFWG meetings, context diagrams, functional flow block diagrams (FFBD), time-line charts, and N^2 charts provide valuable stakeholder input toward establishing a concept consistent with stakeholder needs. Other objectives are:

1. To provide traceability between operational needs and the captured source requirements.
2. To establish a basis for requirements to support the system over its life, such as personnel requirements, support requirements, etc.
3. To establish a basis for verification planning, system-level verification requirements, and any requirements for environmental simulators.
4. To generate operational analysis models to test the validity of external interfaces between the system and its environment, including interactions with external systems.
5. To provide the basis for computation of system capacity, behavior under/overload, and mission-effectiveness calculations.
6. To validate requirements at all levels and to discover implicit requirements overlooked from other sources.

Systems engineering produces a ConOps document early in the requirements definition process. Since a ConOps describes system behavior, a starting point for building up the concept is to begin by identifying outputs generated by external systems (modified as appropriate by passing through the natural system environment), which act as stimuli to the system-of-interest and cause it to take specified actions and produce outputs, which in turn are absorbed by external systems. These single threads of behavior eventually cover every aspect of operational performance, including logistical modes of operation, operation under designated conditions, and behavior required when experiencing mutual interference with multi-object systems.

Aggregation of these single threads of behavior represents a dynamic statement of what the system is required to do. No attempt is made at this stage to define a complete operational concept or to allocate functions to hardware or software elements (this comes later). This ConOps is essentially a functional concept definition and rationale *from the user and customer perspective*. The ConOps is established as follows:

1. Start with the source operational requirements; deduce a set of statements describing the higher-level, mission-oriented system objectives and record them. The following typical source documents serve as inputs for the ConOps:
    - System business case
    - Statement of User Need
    - Technical operational requirements
    - System operational requirements documents
    - Statement of operational objectives
    - SOW
    - Customer Standard Operating Procedures.
2. Review the system objectives with end users and operational personnel and record the conflicts.
3. Define and model the operational boundaries.
4. For each model, generate a context diagram to represent the model boundary.
5. Identify all of the possible types of observable input and output events that can occur between the system and its interacting external systems.
6. If the inputs/outputs are expected to be significantly affected by the environment between the system and the external systems, add concurrent functions to the context diagram to represent these transformations and add input and output events to the database to account for the differences in event timing between when an output is emitted to when an input is received.
7. Record the existence of a system interface between the system and the environment or external system.
8. For each class of interaction between a part of the system and an external system, create a functional flow diagram to model the

sequence of interactions as triggered by the stimuli events generated by the external systems.

9. Add information to trace the function timing from performance requirements and simulate the timing of the functional flow diagrams to confirm operational correctness or to expose dynamic inconsistencies. Review results with users and operational personnel.
10. Develop timelines, approved by end users, to supplement the source requirements.

Draft ConOps are prepared in early project stages during concept definition studies or pre-proposal studies. As concepts evolve, these drafts should be updated for the next project stage. The following measures are often used to gauge the progress and completion of the ConOps activity:

1. Functional Flow Diagrams required and completed
2. Number of system external interfaces
3. Number of scenarios defined
4. Number of unresolved source requirement statements
5. Missing source documents
6. Number of significant dynamic inconsistencies discovered in the source requirements.

4.1.2.6 Generate the System Requirements Document

If one does not already exist, a draft SRD should be generated to represent the customer/user requirements. This is the highest level document to be created by the project. If an SRD already exists, it should be reviewed internally and with the customer to ensure that it is valid, meets the customer needs, and clearly understood by all stakeholders.

4.2 Requirements Analysis Process

4.2.1 Overview

4.2.1.1 Purpose

As stated in ISO/IEC 15288:2008:

> *The purpose of the Requirements Analysis Process is to transform the stakeholder, requirement-driven view of desired services into a technical view of a required product that could deliver those services.*

> *This process builds a representation of a future system that will meet stakeholder requirements and that, as far as constraints permit, does not imply any specific implementation. It results in measurable system requirements that specify, from the supplier's perspective, what characteristics it is to possess and with what magnitude in order to satisfy stakeholder requirements.*[9]

4.2.1.2 Description

System requirements are the foundation of the system definition and form the basis for the architectural design, integration, and verification. Each requirement carries a cost. It is therefore essential that a complete but minimum set of requirements be established from defined stakeholder requirements early in the project life cycle. Changes in requirements later in the development cycle can have a significant cost impact on the project, possibly resulting in cancellation.

Requirements analysis is both iterative and recursive. According to ISO/IEC CD 29148, *Requirements Engineering*[10]:

> *When the application of the same process or set of processes is repeated on the same level of the system, the application is referred to as iterative. Iteration is not only appropriate but also expected. New information is created by the application of a process or set of processes. Typically this information takes the form of questions with respect to requirements, analysed risks or opportunities. Such questions should be resolved before completing the activities of a process or set of processes.*
>
> *When the same set of processes or the same set of process activities are applied to successive levels of system elements within the system structure, the application form is referred to as recursive. The outcomes from one application are used as inputs to the next lower (or higher) system in the system structure to arrive at a more detailed or mature set of outcomes. Such an approach adds value to successive systems in the system structure.*

The output of the process must be compared for traceability to and consistency with the stakeholder requirements, without introducing implementation biases, before being used to drive the Architectural Design Process. The Requirements Analysis Process adds the verification criteria to the defined stakeholder requirements. Figure 4-4 is the context diagram for Requirements Analysis.

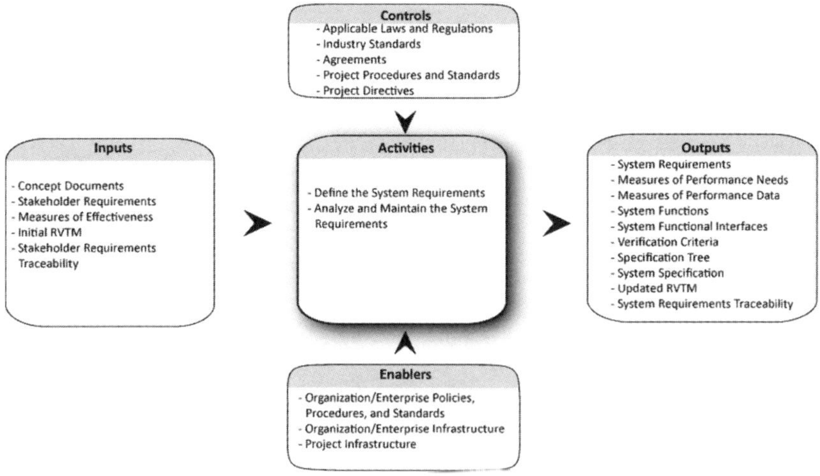

Figure 4-4 Context Diagram for the Requirements Analysis Process

4.2.1.3 Inputs

Inputs to the Requirements Analysis Process include any decisions or data resulting from previous stages of development and also include the following:

- The primary input to the Requirements Analysis Process is the project baseline documented during the Stakeholder Requirements Definition Process:
  - *Concept Documents*
  - *Stakeholder Requirements*
  - *Initial RVTM*
  - *Stakeholder Requirements Traceability.*

This process is governed by the following controls and enablers:

- *Applicable Laws and Regulations*
- *Industry Standards* – relevant industry specifications and standards
- *Agreements* – terms and conditions of the agreements
- *Project Procedures and Standards* – including project plans
- *Project Directives*
- *Organization/Enterprise Policies, Procedures, and Standards* – including guidelines and reporting mechanisms
- *Organization/Enterprise Infrastructure*

- *Project Infrastructure.*

4.2.1.4 Outputs

Outputs of the Requirements Analysis Process are a technical description of characteristics the future system must have to meet Stakeholder Requirements – not a specific solution – that will be evolved in subsequent development processes. These include the following:

- *System Requirements* – The project team derives additional system requirements resulting from analysis of the input Stakeholder Requirements, as required to meet project and design constraints, including:
  - Performance Requirements
  - Functional Requirements
  - Non-Functional Requirements
  - Architectural Constraints.
- *Measures of Performance Needs* – Measures of Performance (MOPs)[11] define the key performance characteristics the system should have when fielded and operated in its intended operating environment (see Section 5.7.2.3).
- *MOP Data* – Data provided to measure the MOPs.
- *System Functions* – Defines the functions the system must perform and defines the functional boundaries for the system to be developed.
- *System Functional Interfaces* – Identifies and documents any interfaces and information exchange requirements with systems external to the functional boundaries.
- *Verification Criteria* – These may specify who will perform verification activities, and the environments of the system-of-interest and enabling systems.
- *Specification Tree* – Based on the evolving system architecture, a specification tree is produced to define the hierarchical representation of the set of specifications for the system under development.
- *System Specification* – Formally documented and approved system requirements may be captured in a document called the System Specification. System specifications are developed through iterative process executions of Requirements Definition, Requirements Analysis, and Architectural Design (see Section 4.3).

- *Updated RVTM.*
- *System Requirements Traceability* – All system requirements should have bi-directional traceability, including to their source, such as the originating stakeholder requirements.

Any decisions are documented in the information repository.

4.2.1.5 Process Activities

The Requirements Analysis Process includes the following activities:

- *Define the System Requirements* – This includes defining and specifying the functional boundary and MOPs. This will specify what the system should be able to do (functional requirements) when fielded and operated in its intended operating environment. The MOPs for the top-level system functional requirements required to satisfy the MOEs[11] are based on the level of risk to achieve the required performance. The system requirements are determined from the following information:
  - *Selected standards* – Identify standards required to meet quality or design considerations imposed as defined stakeholder requirements or derived to meet organization, industry, or domain requirements.
  - *System boundaries* – Clearly identify system elements under design control of the project team and/or organization and expected interactions with systems external to that control boundary as defined in negotiated Interface Control Documents (ICD). After agreement, the ICDs are placed under formal change control.
  - *External interfaces* – Functional and design interfaces to interacting systems, platforms, and/or humans external to the system boundary as negotiated in the ICDs.
  - *System Functions* – Define the functions that the system is to perform. These functions should be kept implementation independent. For more information on approaches for defining and refining system functions see Section 4.12.2.
  - *Utilization environment(s)* – Identify all environmental factors (natural or induced) that may affect system performance, impact human comfort or safety, or cause human error for each of the operational scenarios envisioned for system use.

  - *Life-cycle process requirements* – Conditions or design factors that facilitate and foster efficient and cost-effective life-cycle functions (e.g., Production, Deployment, Transition, Operation, Maintenance, Reengineering/Upgrade, and Disposal).
  - *Design considerations* – Including human systems integration (e.g., manpower, personnel, training, HFE, environment, safety, occupational health, survivability, habitability), system security requirements (e.g., information assurance, anti-tamper provisions), and potential environmental impact.
  - *Design constraints* – Including physical limitations (e.g., weight, form/fit factors), manpower, personnel, and other resource constraints on operation of the system, and defined interfaces with host platforms and interacting systems external to the system boundary, including supply, maintenance, and training infrastructures.
- *Analyze and Maintain the System Requirements*
  - *Define Verification Criteria* – This activity is conducted concurrent with requirements analysis efforts to ensure verifiable requirements.
  - Maintain continuity of configuration control and traceability.

Common approaches and tips:

- IPDTs with acquirer-supplier participation (see Section 5.1.2.3) are an effective practice to bring together the necessary expertise.[12]
- Use failure modes, effects, and criticality analysis (FMECA) or hazard analysis to identify the critical system level requirements.
- Use tools specifically designed to support requirements management.[13]
- Maintain requirements traceability from the beginning of the Requirements Analysis activity.
- Ensure derived requirements are consistent with other requirements or constraints.
- Create templates for constructing requirements statements.[14]

4.2.2 Elaboration

This section elaborates and provides "how-to" information on the requirements analysis and management. Other key information on requirements can be found in ISO/IEC TR 19760, ISO/IEC CD 29148, *Requirements Engineering,*[10] and in EIA 632, *Standard – Processes for Engineering a System*[15] (Requirements 14, 15, and 16, and Annex C3.1 a, b, and c).

4.2.2.1 Requirements Analysis Concepts

Requirements analysis, like the set of SE processes, is an iterative activity in which new requirements are identified and constantly refined as the concept develops and additional details become known. These are analyzed, and deficiencies and cost drivers are identified and reviewed with the customer to establish a requirements baseline for the project.

A second objective of requirements analysis is to provide an understanding of the interactions between the various functions and to obtain a balanced set of requirements based on user objectives. Requirements are not developed in a vacuum. An essential part of the requirements development process is the ConOps, the implicit design concept that accompanies it, and associated demands of relevant technology. Requirements come from a variety of sources, including: the customer/users, regulations/codes, and the corporate entity. Figure 4-5 illustrates this environment.

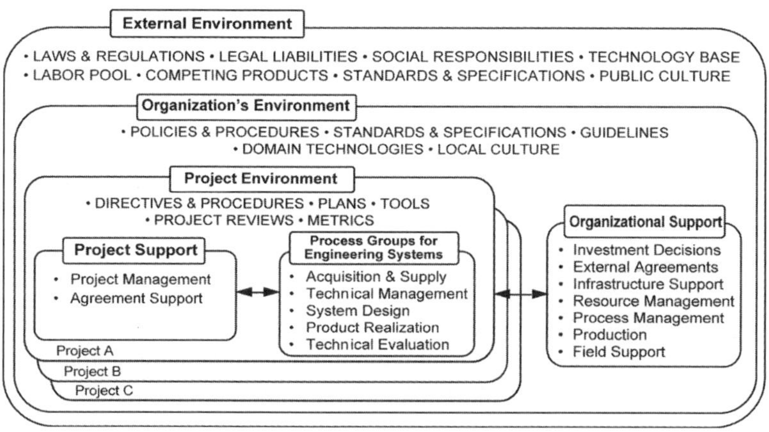

Figure 4-5 Sources of Requirements

This complex process employs performance analysis, trade studies, constraint evaluation and cost/benefit analysis. System requirements cannot be established without determining their impact (achievability) on lower level elements. Therefore, requirements definition and analysis is an iteration and balancing process that works both "top-down" (called allocation and flowdown) and "bottom-up." Once the top-level set of system requirements has been established, it is necessary to allocate and flow them down to successively lower levels. As the allocation and flowdown process is repeated, it is essential that traceability be maintained to ensure that all system level requirements are satisfied in the resulting design. The resulting requirements database (see Section 4.1.2.4) usually contains many attributes for each requirement, and is also used in verification.

4.2.2.2 Characteristics of Good Requirements

In defining requirements, care should be exercised to ensure the requirement is appropriately crafted. The following attributes should be considered for every requirement:

1. *Necessary* – Every requirement generates extra effort in the form of processing, maintenance, and verification. Only necessary requirements should be written. Unnecessary requirements are of two varieties: (1) unnecessary specification of design, which should be left to the discretion of the designer, and (2) a redundant requirement covered in some other combination of requirements.

2. *Implementation Independent* – Customer requirements may be imposed at any level they desire; however, when customer requirements specify design, it should be questioned. A proper requirement should deal with the entity being specified as a "black box" by describing what transformation is to be performed by the "box." The requirement should specify "what" is to be done at that level, not "how" it is to be done at that level.

3. *Clear and Concise* – Requirements must convey what is to be done to the next level of development. Its key purpose is to communicate. Is the requirement clear and concise? Is it possible to interpret the requirement in multiple ways? Are the terms defined? Does the requirement conflict with or contradict another requirement? Each requirement statement should be written to address one and only one concept. Requirements with "and," "or," "commas," or other forms of redundancy can be difficult to verify and should be avoided as it can be difficult to ensure all personnel have a common understanding. Requirements must therefore be written with extreme care. The language used must be clear, exact, and in sufficient detail to meet all reasonable interpretations. A glossary should be used to precisely define often-used terms or terms, such as "process," that could have multiple interpretations.

4. *Complete* – A requirement is a demand on the designer (or implementer) at the next level. Is this requirement at the proper level? When generating requirements, the requirements should be targeted at the next lower level and no lower (except when carrying forward a

legitimate customer design requirement or constraint). The stated requirement should also be complete, measurable and not need further amplification. The stated requirement should provide sufficient capability or characteristics.

5. *Consistent* – In many instances, there are applicable government, industry, and product standards, specifications, and interfaces with which compliance is required. An example might be additional requirements placed on new software developments for possible reusability. Another might be standard test interface connectors for certain product classes.

6. *Achievable* – It is imperative that the implementing designer participate in requirements definition. The designer should have the expertise to assess the achievability of the requirements. In the case of items to be subcontracted, it's important that the expertise of potential subcontractors be represented in the generation of the requirements. Additionally, participation by manufacturing and customers/users can help ensure achievable requirements. IPDTs (see Section 5.1.2.3) and requirements reviews provide mechanisms to achieve these perspectives.

7. *Traceable* – Do all requirements trace to the higher level specification and/or user need? Are there requirements at the higher level not allocated (or allocated, but not picked up)? Those with no allocation may be satisfied at that level of the specification. Requirements with either deficiency should be corrected (see Sections 4.2.2.6 and 4.2.2.7).

8. *Verifiable* – Each requirement must be verified at some level by one of the four standard methods (inspection, analysis, demonstration, or test). A customer may specify, "The range shall be as long as possible." This is a valid but unverifiable requirement. This type of requirement is a signal that a trade study is needed to establish a verifiable maximum range requirement. Each verification requirement should be verifiable by a single method. A requirement requiring multiple methods to verify should be broken into multiple requirements. There is no problem with one method verifying multiple requirements; however, it indicates a potential for consolidating requirements. When the system hierarchy is properly designed, each level of specification has a corresponding level of verification during the verification stage. If element specifications are required to appropriately specify the system, element verification should be performed.

In most writing, it is desirable to substitute words that are more or less synonymous to avoid the constant repetition of a word. However, because few words are exact synonyms, requirements should be written using the same wording with exact meaning established. Care must also be taken in utilizing clear, unambiguous phraseology and punctuation. A misplaced comma can have dramatic ramifications. Verb tense and mood in requirements specifications are very important. The following describes the common use of the forms of the verb "to be" as they apply to specifications:

- "Shall" – Requirements are demands upon the designer or implementer and the resulting product, and the imperative form of the verb, "shall," shall be used in identifying the requirement.
- "Will" – Statement containing "will" identify a future happening. It is used to convey an item of information, explicitly not to be interpreted as a requirement. "The operator will initialize the system by ..." conveys an item of information, not a requirement on the designer of his product. However, some organizations have dropped the distinction between "shall" and "will," and treat either word as a means of stating a requirement.
- "Must" – "Must" is not a requirement, but is considered to be a strong desire by the customer, possibly a goal. "Shall" is preferable to the word "must," and only "shall" statements are verifiable and have to be verified. If both are used in a set of requirements, there is an implication of difference in degree of responsibility upon the implementer.
- Other forms – "To be," "is to be," "are to be," "should," and "should be" are indefinite forms of the verb, and they should be minimized when developing requirements. They are not requirements, but should be considered to be capabilities desired by the customer.

The imperative mood may be used as well in specifying requirements. For example, "The database shall be dumped to magnetic tape every four hours." Requirements done in table format, usually express the processing requirements in the imperative mood. Judicious use of the imperative mood can eliminate many words and enhance the readability of specifications.

The use of certain words should be avoided in requirements in that they convey uncertainty. These include:[10]

- *Superlatives* – such as "best" and "most"
- *Subjective language* – such as "user friendly," "easy to use," and "cost effective"

- *Vague pronouns* – such as "he," "she," "this," "that," "they," "their," "who," "it," and "which"
- *Ambiguous adverbs and adjectives* – such as "almost always," "significant," "minimal," "timely," "real-time," "precisely," "appropriately," "approximately," "various," "multiple," "many," "few," "limited," and "accordingly"
- *Open-ended, non-verifiable terms* – such as "provide support," "but not limited to," and "as a minimum"
- *Comparative phrases* – such as "better than" and "higher quality"
- *Loopholes* – such as "if possible," "as appropriate," and "as applicable"
- *Other indefinites* –such as "etc.," "and so on," "to be determined (TBD)," "to be reviewed (TBR)," and "to be supplied (TBS)." TBD, TBD, and TBS items should be logged and documented in a table at the end of the specification with an assigned person for closure and a due date.

Other characteristics of a set of requirements as a whole should be addressed to ensure that the set of requirements collectively provides for a feasible solution that meets the stakeholder intentions and constraints. These include:[10]

- *Complete* – The set of requirements contains everything pertinent to the definition of system or system element being specified.
- *Consistent* – The set of requirements is not contradictory or duplicated and use the same term for the same item in all requirements.
- *Affordable* – The set of requirements can be satisfied by a solution that is obtainable within life cycle cost, schedule, and technical constraints.
- *Bounded* – The set of requirements maintains the identified scope for the intended solution without increasing beyond what is needed to satisfy user needs.

4.2.2.3 Define Systems Capabilities and Performance Objectives

The concepts of production, deployment, operations, and support serve as an excellent foundation from which systems engineers can discern the required capabilities of the system-of-interest and the relevant performance objectives of the system. Together with identified system constraints, these concepts and capability definitions drive the requirements analysis activities. Typical constraints on the system may include:

- Cost and schedule
- Mandated use of commercial off-the-shelf (COTS) equipment
- Operational environment and use of pre-existing facilities and system elements
- Operational interfaces with other systems or organizations.

As a result of this capability definition activity, a number of performance requirements will be identified. These may include areas such as power, propulsion, communications, data processing, environmental, and human interaction and intervention. In the Maglev Train, for example (see Section 3.6.3), the desire to cover large distances in a brief time established train speed parameters, and the need to carry people suggested safety and maximum noise tolerances. Large systems, such as the Maglev Train, may also justify the development of a high-level system simulation evolved from the system architecture. The simulation should contain sufficient functional elements that the interactions can be properly assessed. The purpose of the simulation is to establish measurable parameters for the functional requirements. This provides the necessary guidance to the designers on the size and capability required of their equipment. In addition, these parameters will be used as an integral part of the Verification Process in establishing the capability of the equipment (and the system) to satisfy user needs.

When time permits, use of an interdisciplinary team to audit the requirements may help ensure the clarity, completeness, and consistency of the set. Such a team can also assess that the requirements are verifiable. Unfortunately, it is possible to write reasonable-sounding requirements that in fact cannot be met, thus requiring waivers throughout the project life cycle to deliver the product.

If there is uncertainty associated with a requirement, it should be identified as needing further attention and even proposed for monitoring as part of the project risk management. The ConOps, for example, can be helpful in identifying adverse consequences of derived requirements:

- Is unnecessary risk being introduced?
- Is the technology producible?
- Are sufficient resources available to move forward?
- Are trade studies needed to determine appropriate ranges of performance?

Resolution of uncertainty should be assigned as a responsibility to an individual, and progress and eventual resolution recorded in the decision database. It is advisable to place assumptions, constraints, uncertainties, and analyses associated with derived requirements in the decision and/or requirements database(s).

4.2.2.4 Define, Derive, and Refine Functional/Performance Requirements

At the beginning of the project, SE is concerned primarily with user requirements analysis – leading to the translation of user needs into basic functions and a quantifiable set of performance requirements that can be translated into design requirements.

Defining, deriving, and refining functional/performance requirements applies to the total system over its life cycle, including its support requirements. These requirements need to be formally documented in a manner that defines the functions and interfaces and characterizes system performance such that they can be flowed down to hardware and software designers.

This is a key SE activity and is the primary focus through System Requirements Review [SRR]. During the requirements analysis, the support from most other disciplines (e.g., software, hardware, manufacturing, quality, verification, specialty, etc.) is necessary to ensure a complete, feasible, and accurate set of requirements that consider all necessary life cycle factors of the system definition. The customer is also a key stakeholder and validates the work as it progresses.

Establishing a total set of system requirements is a complex, time consuming task involving nearly all project areas in an interactive effort. It must be done early, since it forms the basis for all design, manufacturing, verification, operations, maintenance, and disposal efforts, and therefore determines the cost and schedule of the project. The activity is iterative for each stage, with continuous feedback as the level of design detail increases, and flows from the SRD, SOW, Company Policies and Procedures, ConOps Document (or Operations Concept Document), Design Concept, System Hierarchy, and Data Item Description (identifies expected content for specifications). The overall Requirements Analysis Process is shown in Figure 4-6.

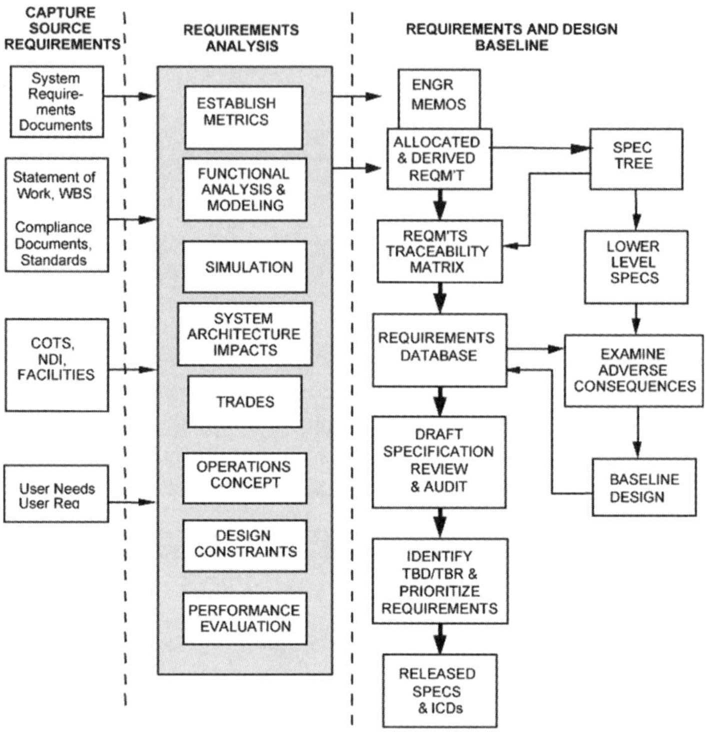

Figure 4-6 Requirements Derivation, Allocation, and Flowdown

The following paragraphs describe the Requirements Analysis Process steps; however, some steps are concurrent and others are not always done in the order shown.

1. Establish constraints on the system. – The starting point for Requirements Analysis is the set of source requirements developed as described in Section 4.1. From the source requirements, Systems Engineers establish constraints on the system including:
    - Cost
    - Schedule
    - Use of COTS equipment
    - Use of Non-Developmental Items (NDI)
    - Use of Existing Facilities
    - Operational Interfaces with other systems or organizations
    - Operational environment.

As a result of this activity, a number of functional and performance requirements are identified.

2. Examine and characterize the mission in measurable requirement categories, such as: quantity, quality, coverage, timeliness, and availability. – Actual systems have many measurables under each attribute, as well as additional attributes, which may include communications, command and control, and security.

3. Using detailed functional analysis (see Section 4.12.2), extract new functional requirements, particularly those required to support the mission. – This should include items such as power, propulsion, communications, data processing, attitude control or pointing, commanding, and human interaction and intervention. This will eventually result in the conversion from mission parameters (e.g., customers supported per node) into parameters that the hardware and software designers can relate to, for example Effective Radiated Power and Received Signal Strength Intensity. The ConOps is a rich source of information for this analysis. Functional decomposition tools, such as functional block diagrams, functional flow diagrams, time lines, and control/data flow diagrams, are also useful in developing requirements. As requirements are derived, the analysis that leads to their definition must be documented and placed into the requirements database.

   Quality Function Deployment (QFD) is a useful technique, particularly where the "voice of the customer" is not clear (see Figure 4-7). It provides a fast way to translate customer requirements into specifications and systematically flowdown the requirements to lower levels of design, parts, manufacturing, and production.

   The shaded Relationship Matrix shows the correlation between features and requirements. Two concentric circles (double circle) are used to indicate a strong correlation between the feature and the requirement. A modest contribution is indicated by a single circle. A blank column indicates an unnecessary feature relative to the listed requirements. Similarly, a blank row indicates an unaddressed requirement. Other useful methods include: functional decomposition using a system hierarchy, FFBDs, time lines, control/data flow diagrams, trade studies, and requirements allocation sheets.

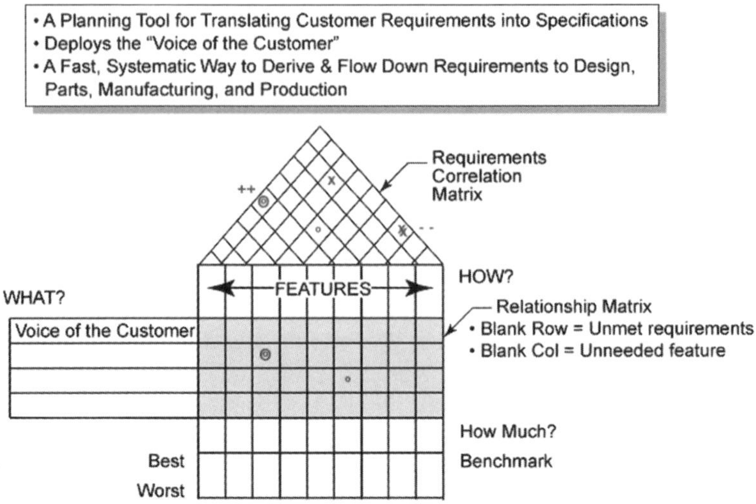

Figure 4-7 Quality Function Deployment (QFD): The House of Quality

4. As discussed above, larger systems may require a high-level system simulation evolved from the system architecture. The simulation will be used to quickly examine a range of sizes and parameters, not just a "Point Design," to ensure that the "best" solution is obtained –the system is the proper size throughout, with no choke points. A number of scenarios should be run using scenarios extracted from the ConOps (with inputs based on system requirements) to exercise the system over the possible range of mission activities. Monte Carlo runs may be made to get averages and probability distributions and to help establish (or verify) timeliness requirements. In addition to examining nominal conditions, non-nominal runs should also be made to establish system reactions or breakage when exposed to extraordinary (out-of-spec) conditions.

5. Examine any adverse consequences introduced by deriving and incorporating requirements. – For example:

    - Is unnecessary risk being introduced?
    - Is the system cost within budget limitations and the budget profile?
    - Will the technology be ready for production?
    - Are sufficient resources available for production and operation?

- Is the schedule realistic and achievable (be sure to consider downstream activities such as design and verification associated with the requirements)?

6. Where existing user requirements cannot be confirmed, perform trade studies to determine more appropriate requirements, and achieve the best-balanced performance at minimum cost. Where critical resources (e.g., weight, power, memory, and throughput) must be allocated, trade studies may be required to determine the proper allocation.
7. Incorporate revised and derived requirements and parameters resulting from the Requirements Analysis Process into the requirements database and maintain traceability to source requirements.
8. Prepare and submit the specification documents (see Sections 4.2.2.6 and 4.2.2.7) to all organizations for review. – Upon approval, the documents are entered into the formal release system, and maintained under configuration management control. Any further changes will require Configuration Control Board (CCB) approval.

The result of the Requirements Analysis Process should be a baseline set of complete, accurate, non-ambiguous system requirements, recorded in the requirements database, accessible to all parties, and documented in an approved, released System Specification. The following measures are often used to gauge the progress and completion of this requirements analysis activity:

1. Number or percent of requirements defined, allocated, and traced
2. Time to issue draft
3. Number of meetings held
4. Number and trends of TBD, TBR, and TBS requirements
5. Number of requirement issues identified (e.g. requirements not stated in a verifiable way)
6. Number and frequency of changes (additions, modifications, and deletions).

4.2.2.5 Define Other Non-Functional Requirements

The concept documents will also suggest requirements that are not directly related to the primary capability provided by the system-of-interest, such as availability, supportability, security, and training. For example, the Øresund Bridge case (see Section 3.6.2) illustrated the avoidance of negative environmental impact by establishing constraints on the construction practices.

Addressing non-functional requirements from the earliest stages is a good way to ensure that they are not forgotten and that they are satisfied.

4.2.2.6 Develop Specification Trees and Specifications

In practice, requirements engineering is not just a front-end to the system development process but a complex communication and negotiation process involving the parties that will use the system (i.e., the customers), the parties that will provide parts or all of the system (i.e., the developers and vendors), and the parties that will verify the system (i.e., the verification group[s]). SE acts as the translator in this communications process with the specifications being the key written embodiment of this communication. Some of the major challenges in performing this requirements engineering task are as follows:

- An envisioned system is seldom, if ever, designed to work totally independent of the other systems in the customer's environment. This means that the environment in which the system is to operate must be known and documented as thoroughly as the system itself.
- COTS solutions play a major role in defining the system. While requirements are supposed to be independent of solution, being able to achieve an implementable solution within the resource constraints available is the primary requirement.
- Every aspect of an envisioned system's function and performance cannot practically be specified. Thus, a level of requirement specification must be established that represents a cost-effective balance between the cost of generating, implementing, and verifying requirements versus the risk of not getting a system that meets customers' expectations. In each case, the cost of non-performance is a major driver.

The SE process is a bridging process translating an identified need into a system solution composed of specified implementable hardware and software elements. The process is very much a communication process with all the potential flaws of any communication, plus the added uncertainty of the customer's real desires and the risks associated with achieving an implementation.

For complex systems, the definition/design process is successively applied through multiple hierarchical iterations down to the level of hardware and software CI definition. The objective is to create a specification baseline for each of the CIs at a particular level of system design (e.g., hardware, software, and operations) and place these specifications in a flowdown hierarchy. This will allow the further definition of each CI to proceed independently, in parallel with all the others, while maintaining requirements traceability and compatibility of all items that make up the system. The roadmap and hierarchical representation of the set of specifications for the system under development is the Specification Tree.

The Specification Tree and Specification activity is led by SE, with support from design engineering and the supporting disciplines. SE creates the Specification Tree and the outlines for each of the specifications, crafts the requirements, and establishes traceability. SE also ensures that the supporting disciplines are present when needed and actively participate, and ensures that their contributions are coordinated and integrated.

Design engineering provides technical definition data for derived requirements, and documents design decisions. Supporting disciplines monitor implementation of requirements in each specialty area, identify requirements, and review the results of the requirement definition process. The output of this effort is a set of requirements statements, which are placed in the system and CI specifications. Draft specifications are generated by the requirements database, and distributed to reviewers. The copies are then returned with comments as appropriate, to the author. When all comments are resolved, the document is formally released. The Requirements Database tool should generate the specification directly from the database without manual intervention, thereby preserving the integrity of the database. Note: specifications at one level represent requirements to the levels below it.

The following paragraphs describe the steps of the Specification Tree and Specification Development activity.

1. Derive the Specification Tree from the system architecture configuration. – As discussed in Section 2.4, *The Hierarchy Within A System*, the system hierarchy should be balanced with appropriate fan-out and span of control. A level of design with too few entities likely does not have distinct design activity, and both design and verification activities contain redundancy. Figure 4-8 shows a typical specification tree corresponding to a balanced system hierarchy. As shown, hardware CIs (HWCI) can have computer software CIs (CSCI) subordinate to them. For example, a display screen on a mobile phone is hardware, but it must have embedded software to operate. Also, the operating system in the mobile phone is a computer software CI, but software defines subordinate hardware requirements to meet the higher-level software requirements (e.g., capacity and speed).

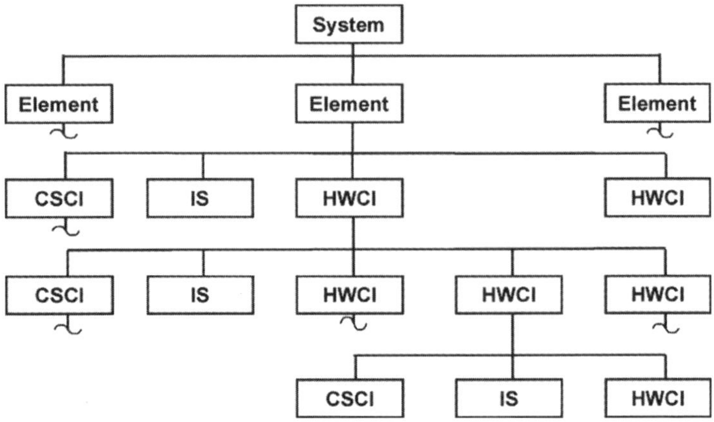

HWCI	Hardware Configuration Item
CSCI	Computer Software Configuration Item
IS	Interface Specification

Figure 4-8 Example Project Specification Tree, also known as a Product Breakdown Structure (PBS)

Developing the specification tree (such as the one shown here) is one element of system design whereby the system is decomposed into its constituent parts. This activity has major ramifications on the development of the system in that it essentially determines the items to be purchased versus those to be developed and establishes the framework for the integration and verification program. The objective in the design is to achieve the most cost-effective solution to the customer's requirements with all factors considered. Generally, this is achieved by identifying existing or implementation units as early as possible in the tree development. At each element or node of the tree a specification is written, and later on in the project, a corresponding individual verification will be performed. When identifying elements, it is useful to consider the element both from a design and a verification perspective. The element should be appropriate from both perspectives. The design and development methods described in the earlier sections of this handbook apply to this activity. As for the actual generation of the Specification Tree and Specifications, templates and previously completed specifications are useful starting points for document generation.

2. Create an outline for each specification in the Specification Tree using a standard specification template and the definition of the CI. – Specification outlines or templates may be obtained from several

sources. The most useful and commonly used are previous similar specifications prepared by your organization. These are often the source of useful material, parts of which can be used with minimal modification. Standard formats and IEEE formats recommended for system, hardware, and software specifications are also available.

3. Craft requirements for each specification, completing all flowdown and accommodating derived requirements emerging from the definitions of each CI. – A specification represents a design entity and a verification entity. The specification should represent appropriate complexity from both the design and the verification perspective. Many factors contribute to the appropriate selection of elements. However, as a measure of complexity, a requirements specification should not have too many or too few requirements. As a "rule of thumb" 50 to 250 functional/performance requirements in a specification is appropriate. Requirements in the physical or environmental areas would be in addition to the functional/performance variety.

The result of this activity is the Specification Tree and the set of specifications for all of the CIs that implement the system. As such, completion of the this activity typically occurs when (1) all specifications have been identified and located on the Specification Tree, and (2) each specification is adequate to proceed with the next stage of development or procurement. The following measures are often used to gauge the progress and completion of this activity.

For the Specification Tree:

1. Its completeness as measured by the inclusion of all items required in the system
2. Its balance as determined by its span of control and fan-out from each entity.

For the Specifications:

1. Number of TBDs and TBRs in specifications (goal is zero)
2. Number of requirements in the specification (50 to 250 functional/performance requirements is the ideal range)
3. Stability of the requirements as the development progresses.

4.2.2.7 *Allocate Requirements and Establish Traceability*

Traceability is not an end goal in and of itself but, rather, a tool that can be used to:

1. Improve the integrity and accuracy of all requirements, from the system level all the way down to the lowest CI
2. Allow tracking of the requirements development and allocation and generating overall measures
3. Support easier maintenance and change implementation of the system in the future.

The initial definition of system requirements from the source documents is completed using a combination of graphical functional analysis tools and simulations. As the requirements are developed, a design concept and a ConOps are developed concurrently. Inputs to this process include the Specification Tree, the SRD, and the System Specification. The essential point is that every requirement at every level should have a clear definition of its source and why it is needed.

Derived requirements (i.e., those requirements that are indirectly traced to higher-level requirements; see definition in RTCA/DO-178B) must be documented and their justification given. Other requirements (i.e., the primary requirements) are justified by virtue of the fact that they trace directly to a higher-level requirement. For example, every requirement in the System Specification should be traceable to the stakeholder requirements and the SRD.

Bi-directional traceability is top-down and to V&V plans and procedures for specifications (CI and interface), and should include traceability to the verification program (e.g., plans, procedures, test cases, and reports) to provide closed-loop verification. Traceability should be maintained throughout all levels of documentation, as follows:

1. Allocate all system requirements to hardware, software, or manual operations, facilities, interfaces, services, or others as required
2. Ensure that all functional and performance requirements or design constraints, either derived from or flowed down directly to a lower system architecture element, actually have been allocated to that element
3. Ensure that traceability of requirements from source documentation is maintained through the project's life until the verification program is completed and the system is accepted by the customer

4. Ensure that the history of each requirement on the system is maintained and is retrievable.

The following activities provide additional guidance for establishing and maintaining requirement traceability:

1. While requirements can be traced manually on small projects, such an approach is generally not considered cost-effective, particularly with the proliferation of requirements management tools. A requirements traceability tool that augments the requirements database should be accessible to and usable by all technical personnel on the project. This includes subcontractors who are preparing specifications and verification data. The tool should generate the following directly from the database:

    a. Requirements Statements with PUIDs

    b. RVTM –a list of requirements, their verification attributes, and their traces

    c. Requirements Traceability Matrices (RTM) – list of requirements and their traces

    d. Lists of TBD, TBR, and TBS issues

    e. Specifications

    f. Requirements measures (e.g. requirements stability).

    The tool must also have configuration management capability to provide traceability of requirements changes and ensure that only properly authorized changes are made.

2. Each requirement must be traceable using a PUID. The PUID is an alphanumeric assigned to each requirement. The alphanumerics employed are similar to acronyms to provide an easily recognizable identification of the specification for the requirements statements. This is particularly useful when requirements statements are extracted from many specifications as part of the audit process. The numeric portion is assigned within individual documents.

3. The specification tree (see Section 4.2.2.6) provides the framework for parent-child vertical traceability (tree-down or tree-up) used for specifications. For interface documents, such as ICDs, the traceability is, in some cases, over several levels. Thus, the specification tree does not adequately portray interface traceability. Therefore, the requirements tool must have capability for bi-directional traceability top-down and to verification plans and procedures.

4. The functions and sub-functions for which each system area is responsible, and the top level system requirements associated with those functions, must be identified and documented in the traceability tool. The process for identifying sub-functions and allocating requirements to those functions is described in Section 4.12.2, *Functions-Based Systems Engineering Method*. As with requirements, a PUID is assigned to each of the functions (system actions) and sub-functions, and the functional/performance requirements to be associated with each function are identified and captured in the database. Finally, each function and sub-function is allocated to an element in the system architecture, and the related information is captured in the database.

5. The most difficult part of requirements flowdown can be the derivation of new requirements, which often involves a change in the parameters as appropriate to the level in the hierarchy (e.g., targets per sq. mi – a system parameter – has little meaning to the hardware designer). Each branch of the system tree may be decomposed to a different level, depending on the complexity and the system acquisition approach (e.g., make or code everything or use suppliers or subcontractors for one or more branches). At the lowest CI level to be defined, the parameters specified must be relevant to that particular item (hardware, software, or composite) and provide adequate direction to the designer. The derivation and decomposition activity is repeated at each level of the requirements hierarchy until the appropriate level is reached.

6. The specifications should be reviewed and audited as they are produced to verify that the allocation activity is correct and complete. As such, the Requirements Database should be capable of generating status and audit reports that contain the flowdown of requirements statements. These reports allow Systems Engineers to identify proposed corrections and changes, and process them through the proper approval channels.

   As the system life cycle proceeds, increasing effort will be directed toward verification that the demonstrated capability of the system meets its requirements as expressed in specifications. The requirements/traceability database plays a major role in this by incorporating the verification data in its attribute files, either directly or by pointer to other databases where the data are located. Status reports on verification progress, progress in eliminating undefined requirements (TBD/TBR/TBS), and requirements changes can be obtained by sorting the appropriate attribute listings.

7. Once allocations are verified, RTMs are generated directly from the database and maintained under configuration management control. These matrices are used as part of the audit process.

Traceability is achieved when all requirements at a particular level of the system hierarchy have been placed in the database and traced up and down, as appropriate. A complete set of allocated requirements should be found in specifications, with an RTM. The following measures are often used to gauge the progress and completion of the allocation and traceability activity:

1. Number and trends of requirements in the database
2. Number of TBD, TBR, and TBS requirements
3. Number (or percent) of system requirements traceable to each lower level and number (percent) of lower level requirements traceable back to system requirements.

4.2.2.8 Generate the System Specification

The system specification is a baseline set of complete, accurate, non-ambiguous system requirements, recorded in the requirements database and accessible to all parties. To be non-ambiguous, requirements must be broken down into constituent parts in a traceable hierarchy such that each individual requirement statement is:

- Clear, unique, consistent, stand-alone (not grouped), and verifiable
- Traceable to an identified source requirement
- Not redundant, nor in conflict with, any other known requirement
- Not biased by any particular implementation.

These objectives may not be achievable using source requirements. Often requirements analysis is required to resolve potential conflicts and redundancies and to further decompose requirements so that each applies only to a single system function. Use of an automated requirements database will greatly facilitate this effort, but is not explicitly required.

During the Requirements Analysis Process, it is often necessary to generate a "snapshot" report of clarified system requirements. To aid this process, it may be desirable to create a set of clarified requirement objects in the requirements database with information providing traceability from their corresponding originating requirement. Clarified requirements may be grouped as functional, performance, constraining, and non-functional.

4.3 Architectural Design Process

4.3.1 Overview

4.3.1.1 Purpose

As stated in ISO/IEC 15288:2008:

> *The purpose of the Architectural Design Process is to synthesize a solution that satisfies system requirements.*
>
> *This process encapsulates and defines areas of solution expressed as a set of separate problems of manageable, conceptual and, ultimately, realizable proportions. It identifies and explores one or more implementation strategies at a level of detail consistent with the system's technical and commercial requirements and risks. From this, an architectural design solution is defined in terms of the requirements for the set of system elements from which the system is configured. The specified design requirements resulting from this process are the basis for verifying the realized system and for devising an assembly and verification strategy.*[16]

4.3.1.2 Description

The Architectural Design Process is iterative and requires the participation of systems engineers joined by relevant specialists in the system domain. When alternative solutions present themselves, technical analysis and decisions are made as part of this process to identify a set of system elements. Figure 4-9 is the context diagram for the Architectural Design Process.

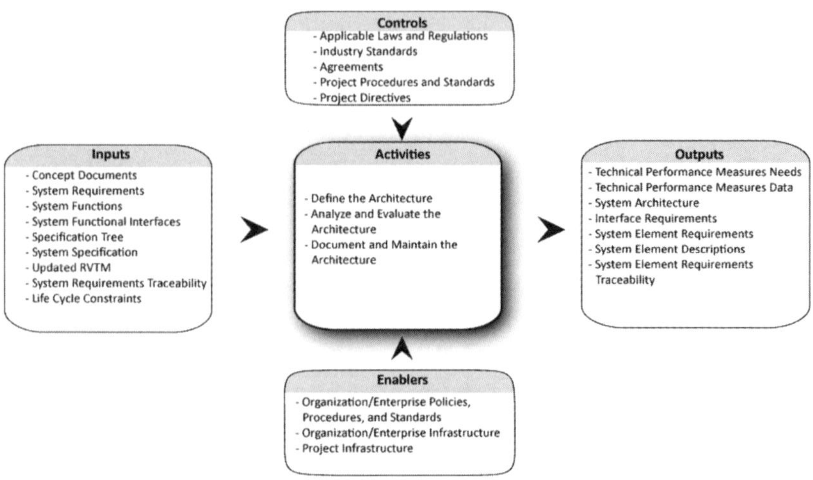

Figure 4-9 Context Diagram for the Architectural Design Process

4.3.1.3 Inputs

Inputs to the Architectural Design Process include the following:

- The primary input to the Architectural Design Process is the project baseline documented during the Requirements Analysis and Stakeholder Requirements Definition Processes:
  - *Concept Documents*
  - *System Requirements*
  - *System Functions*
  - *System Functional Interfaces*
  - *Specification Tree*
  - *System Specification*
  - *Updated RVTM*
  - *System Requirements Traceability.*
- *Life Cycle Constraints* – Life-cycle related items that may influence the system design and architecture. A constraint is a limitation or implied requirement that constrains the design solution or implementation of the SE process and is not changeable by the enterprise.[17] Constraints include:
  - Implementation Constraints on Design
  - Integration Constraints on Design
  - Verification Constraints on Design
  - Transition Constraints on Design
  - Validation Constraints on Design
  - Operation Constraints on Design
  - Maintenance Constraints on Design
  - Disposal Constraints on Design.

This process is governed by the following controls and enablers:

- *Applicable Laws and Regulations*
- *Industry Standards* – relevant industry specifications and standards
- *Agreements* – terms and conditions of the agreements
- *Project Procedures and Standards* – including project plans

- *Project Directives*
- *Organization/Enterprise Policies, Procedures, and Standards* – including guidelines and reporting mechanisms
- *Organization/Enterprise Infrastructure*
- *Project Infrastructure*.

4.3.1.4 Outputs

The result of the Architectural Design Process is an architectural design that is placed under configuration management. This baseline includes:

- *TPM Needs* –TPMs[11] are measures tracked to influence the system design (see Section 5.7.2.4)
- *TPM Data* – Data provided to measure the TPMs
- *System Architecture Description* – Description of the system architecture, typically presented in a set of architectural views, along with documented justification for concept selections
- *Interface Requirements* – Interface requirements supporting a plan for system integration and verification strategy
- *System Element Requirements* – Allocated and derived requirements assigned to system elements and documented in a traceability matrix
- *System Element Descriptions* – Detailed system element descriptions
- *System Element Requirements Traceability* – All system element requirements should have bi-directional traceability, including to their source, such as the originating system requirements.

4.3.1.5 Process Activities

The Architectural Design Process includes the following activities:

- *Define the Architecture*
  - Define a consistent logical architecture – capture the logical sequencing and interaction of system functions or logical elements.
  - Partition system requirements and allocate them to system elements with associated performance requirements – Evaluate COTS solutions that already exist.
  - Identify interfaces and interactions between system elements (including human elements of the system) and with external and enabling systems.

- Define V&V Criteria for the system elements.

- *Analyze and Evaluate the Architecture*
  - Evaluate COTS elements for compatibility with the design.
  - Evaluate alternative design solutions (see Section 5.3) using the selection criteria defined in Section 4.3.2.2.
  - Support definition of the system integration strategy and plan (to include human systems integration [HSI]).
- *Document and Maintain the Architecture*
  - Document and maintain the architectural design and relevant decisions made to reach agreement on the baseline design.
  - Establish and maintain the traceability between requirements and system elements.

Common approaches and tips:

- Modeling techniques, such as SysML™ (see Section 4.12.3), are useful in deriving a logical architecture.
- IPDTs (see Section 5.1.2.3) facilitate in-depth analysis and review; working together as a team helps break down communications silos and facilitates informed decision-making.
- Architecture and Design Patterns can be useful for establishing a system framework.
- System elements can be developed in a top-down partitioning exercise that allocates the functional and non-functional elements to physical or virtual system elements. Ideally, interface requirements between these system elements are minimized. At the same time, COTS or previously developed system elements are considered within the constraints of the contracting strategy.
- During this process, consider emergent properties, feature interactions, and human-system interactions.[17]

4.3.2 Elaboration

4.3.2.1 Architectural Design Concepts

In his book *Systems Analysis, Design and Development*, Charles Wasson states, "System, product, or service architectures depict the summation of a system's entities and capabilities at levels of abstraction that support all stages of deployment, operations, and support."[19] As such, developing the system architecture is one of the most important responsibilities of the systems engineer. The overall objective is to create a System Architecture (defined as the selection of the types of system elements, their characteristics, and their arrangement) that meets the following criteria:

1. Satisfies the requirements (including external interfaces)
2. Implements the functional architecture
3. Is acceptably close to the true optimum within the constraints of time, budget, available knowledge and skills, and other resources
4. Is consistent with the technical maturity and acceptable risks of available elements.

Good SE practice considers multiple candidate system solutions, including multiple views of a single architecture (such as advocated by Zackman[20,21] and required by DODAF and MODAF) as well as multiple competing architectures. Figure 4-10 illustrates this concept by considering two architectural alternatives that are significantly different in their approach to meeting stakeholder requirements for intercontinental telephone communication.

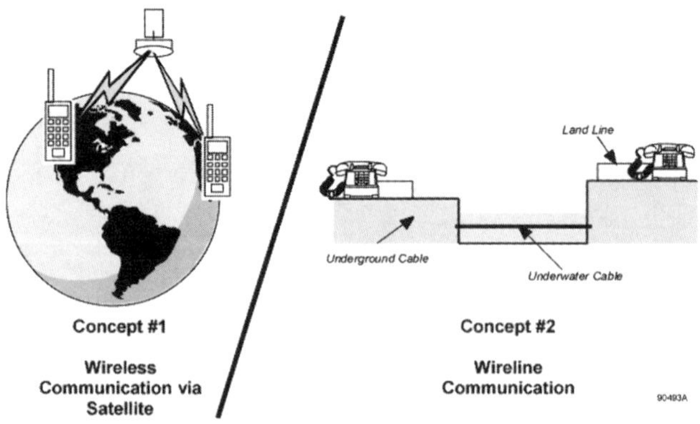

Figure 4-10 Example of Alternative Architectural Concepts

This illustration correctly implies that there is no unique solution to satisfying user requirements. Hence, the system architecture is critical because it provides the framework for system development.

Creating an effective system architecture draws on the experience, intuition, and good judgment of the team to devise an appropriate solution. As Rechtin and Maier define it, systems architecting builds on four methodologies:[22]

- Normative (solution-based), such as building codes and communication standards.

- Rational (method-based), such as systems analysis and engineering.
- Participative (stakeholder-based), such as concurrent engineering and brainstorming.
- Heuristic (lessons-learned), such as "Simplify. Simplify. Simplify."

Because systems architecting is a creative process, and because intuition and experience play such an important role, the systems engineer must pay attention to situations where past experience and intuition have been a handicap. *The Innovator's Dilemma*,[23] by Clayton Christensen, sets forth the benefit of making creative use of past experience with an example from the computer hard disk drive industry, in which experience had been a key factor to the growth of companies that dominated the hard disk market. Christensen highlights the transition difficulties faced by companies making large disks and their inability to capture any market share when the industry moved to smaller-size disks. According to Christensen, no manufacturer made a successful transition from a 14-inch (35.6 cm) disk to the 8-inch (20.3 cm) disk; consequently, a whole new set of companies dominated that market. This was repeated as the sizes dropped from 8 to 5-¼ to 3-½ to 2-½ to 1.8 inches. This sequence started in the 1980s and is continuing today with the introduction and evolution flash drives. In each of these transitions, the established companies lost out, in part because their established user base was locked-in to the older architecture, and in part because their entire organization from SE to marketing to manufacturing to executive management was unable to see the new vision.

Architectural design is strongly focused on analysis of alternatives and is part of the overall system definition that includes Stakeholder Requirements Definition (see Section 4.1), Requirements Analysis (see Section 4.1.2.6), and Decision Management (see Section 5.3) processes. An initial set of functions is defined to carry out the system's overall mission. Requirements are derived to quantify how well the functions must be performed and to impose constraints on the system. An architecture is then chosen to implement the functions and satisfy the requirements and constraints. The realities of a practical architecture may reveal a need for additional functional and performance requirements, corresponding to architecture features necessary for wholeness of the design, but not invoked by the original set of functions. Similarly, the initial functional and performance requirements may prove infeasible or too costly with any realizable architecture. Consequently, the process involves the mutual, iterative adjustment of functions, requirements, and architecture until a compatible set has been discovered. The Architecture Design Process flows as shown in Figure 4-11.

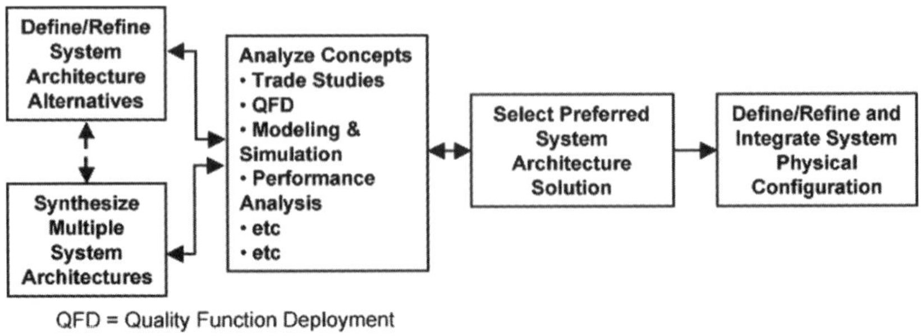

Figure 4-11 System Architecture Synthesis Process Flow

The Architectural Design Process is essentially a tradeoff, performed at a grand scale, leading to a selected system architecture baseline as the final output. The objective is to select the best alternative from among a set of System Architecture candidates, which have been constructed in a manner that ensures (with reasonable certainty) that one of the candidates is acceptably close to the true (usually unknowable and unattainable) optimum.

The limitations of text necessitate a sequential description of process activities, although in practice, the process usually proceeds in a highly-interactive, parallel manner with considerable iteration. In addition, for clarity and completeness, a rather formalized description of each activity is provided in the sections that follow. For large projects, with numerous participating organizations at separate locations, the process may require a high level of formality and discipline for coordination and concurrence, whereas a small unified team can operate in a much more informal manner. ISO/IEC 42010, *Architecture Description*, contains a valid definition for an architecture description.

4.3.2.2 Define Selection Criteria

Selection criteria are the quantifiable consequences of system implementation and operation. They are derived from the system requirements, ConOps, and functional architecture, and from programmatic considerations, such as available resources (financial and otherwise), acceptable risk, political considerations, and the updated business case for the system. This activity is conducted by SE with involvement from specialists, as necessary, to support the definition of selection criteria and the modeling and analysis used to make the selection. Selection criteria include:

1. Measures of the system's ability to fulfill its mission as defined by the requirements

2. Ability to operate within resource constraints

3. Accommodation of interfaces
4. Ability to adapt to projected future needs and interoperating systems (i.e., system robustness)
5. Costs (economic and otherwise) of implementing and operating the system over its entire life cycle
6. Side effects, both positive and adverse, associated with particular architecture options
7. Measures of risk
8. Measures of quality factors
9. Measures of subjective factors that make the system more or less acceptable to customers, users, or clients (e. g., aesthetic characteristics).

4.3.2.3 Define/Refine System Element Alternatives

The purpose of this activity is to identify and refine a set of element options (including all the hardware, software, information, procedures, interfaces, and people that make up the system), one level down from the top of the system hierarchy. These options constitute a set of building blocks from which System Architecture options will be assembled and, in general, should satisfy the following criteria:

- With reasonable certainty, spans the region of design space that contains the optimum
- Supports analysis that efficiently closes on the optimum
- Contains all relevant design features necessary to provide a firm baseline for the subsequent round of system definition at the next level of detail.

The system element definition activity is led by SE with major support from hardware and software design and operations. Other engineering support disciplines participate, as appropriate, to support the definition of system element options, in particular, to anticipate issues that will become important later in the system definition and design process. Within the framework of concurrent engineering, all disciplines are kept informed as the process unfolds. Specific comments may be solicited, and any discipline is free to contribute at any time. The objective is to include all disciplines in identifying design drivers, defining selection criteria, and detecting show-stoppers. In addition, participation in these earlier stages of system definition lays the groundwork for knowledgeable contributions later in the process.

The following steps, although listed in sequential order, are highly interactive and usually evolve in a parallel and iterative manner.

1. Create a list of the elements that will make up the system. – These may be derived from the functional requirements or from a decomposition of the existing system architecture. Additional sources for identifying system elements include:

    - Business case for the system, including cost and schedule goals or hard limits
    - Requirements and Life-Cycle Operational Concept, Functional Architecture
    - Technology (available and emerging), and technical constraints
    - Examples of existing systems or elements which perform similar functions
    - System-dependent CONOPS unique to each concept.

    The initial version of this list of elements may be considered preliminary, and will mature as the process is iterated.

2. Identify a set of option descriptors for each element in the list. – These are the definitive attributes (design features and parameters) that distinguish one element option from another. The descriptors are the minimal set of significant element characteristics that allows a unique identification for every element choice in the design space.

3. Define the envelope of design space (range of design features and parameter values) that is to be scanned.

4. Develop a process to generate a range of element options, providing both diversity of element types and range of design parameters within a given type. – This includes demonstrating that the range of element options created is both exhaustive and lean:

    - *Exhaustive* – no good options have been left out and the optimum is somewhere within the envelope of options under consideration (the definition of optimum includes satisfaction of all requirements and factors, such as acceptable design maturity, compatibility with the development schedule, minimum cost, acceptable risk, etc.).
    - *Lean* – the number of options to be analyzed is small enough to support efficient selection and closure on the optimum.

The range of element options in the set may be defined by either (1) expanding a range of various types of elements representing diverse approaches to implementing the system functions (e.g., if the function is communications, the types of elements might include microwave relay, satellite link, or fiber optics), or (2) considering variations of design parameters within any given element type (e.g., number and thrust level of chambers needed to produce a required total thrust for a launch vehicle).

5. Borrow from similar existing systems or create new element options through application of the appropriate structured creativity methods. – Some useful methods include brainstorming, morphological analysis, synectics, (see Adams[24] and other references on structured creativity), literature search, surveys, inventory of existing concepts, and vendor inquiries. Any element previously defined or inferred by the Requirements Analysis, ConOps Concept, or Functional Analysis must be included.

6. Generate a set of element options that populates the design space envelope. – In general, the options selected should satisfy all requirements, but it is useful to include some that may challenge the requirements in ways leading to a better system concept. This includes relaxing requirements of marginal utility or are costly to implement, or extending requirements where added capability can be purchased cheaply.

7. Develop the attendant data describing each element option and its interfaces with other elements, as needed, to support the selection process and subsequent system definition activity. – These data should include estimates for cost, performance, development time, and risk descriptions for each option.

The result of performing this activity is a set of element options with descriptive and supporting documentation that provides:

1. Set of descriptors that define the dimensions of the design space.

2. Set of element options, each characterized by a description of its salient features, parameter values, and interactions with other elements as necessary to characterize it for analysis and as a potential baseline CI. This can take the form of diagrams, schematics, concept drawings, tabular data, and narrative.

3. Supporting documentation of the rationale that justifies the selection of the descriptors, the design space envelope, and the menu of element options.

4. Identification of design drivers (i.e., a limited set of top-level parameters that dominate definition of the design), definition of selection criteria related to elements that will be used in the evaluation process, and detection of issues that contain possible show-stoppers.

5. Documented assurance, with reasonable certainty for the set of options as a whole, that a basis has been established for efficient selection of the optimum architecture. This should include assurance that the options selected will meet the requirements, that the optimum is somewhere within the range of options to be analyzed, that the optimum can be found quickly and with reasonable certainty, and that the descriptive data (features and parameters) are adequate to support subsequent system definition work.

The following measures are used to gauge the progress and completion of the system element definition activity:

1. Technical performance, schedule spans, costs, and risk estimates for each alternative

2. Evidence that each alternative is consistent with the business case for the system.

Quality Functional Deployment (see Section 4.2.2.4, Figure 4-7) provides a framework to organize the data and verify the completeness of the analysis.

4.3.2.4 Synthesize Multiple System Architectures

Synthesizing a system architecture consists of (1) selecting the types of system elements that comprise the system (created by the activity described above), (2) assessing their characteristics, and (3) determining an effective arrangement that fits within the design space of possible System Architecture arrangements of those elements. The objective is to provide a set of candidate System Architecture options from which a final optimized and robust System Architecture will be selected or evolve.

This activity is led by SE with major support from teams responsible for hardware and software design, operations, human factors, producibility, logistics, safety, and others. Other engineering support disciplines participate, as appropriate, to support the definition of system element characteristics and creation of arrangement options, and in particular, to anticipate issues that will become important later in the system definition and design process. Key tasks associated with this activity are as follows:

1. Assemble candidate System Architectures. –
    a. Examine the System Architecture of existing systems that perform similar functions and adopt, in existing or modified form, any architectures that appear suitable.
    b. Generate system architecture options by combining elements from the set of element options. For each top-level system function, identify a range of means by which is the architecture will be implemented (choice of implementing element type or design), and build a set of integrated system concepts that incorporate all the element choices. The methodology should rule out absurd or obviously non-optimal combinations of elements, and seek particularly appealing new combinations of the elements. Some useful methods include brainstorming, morphological analysis, synectics, literature search, surveys, inventory of existing concepts, and vendor inquiries.
2. Verify that the resulting System Architecture options meet the following criteria:
    a. Perform all the functions of the system
    b. Capable of meeting requirements
    c. Satisfies all constraints
    d. Resource usage is within acceptable limits
    e. Elements are compatible
    f. All interfaces are satisfied.
3. Ensure in-process validation by involving the customer or user in this process.
4. Screen the set of System Architecture options generated so far, retaining only a reasonable number of the best. – Modify the options as necessary to distribute them with reasonable separation throughout the most promising region of the design space. If some promising regions of design space are poorly represented, create more options to fill the void.

The INCOSE website contains a list of current references of applicable tools for accomplishing this activity. Quality Functional Deployment (see Section 4.2.2.4, Figure 4-7) provides a framework to organize the data and verify the completeness of the analysis. Others include but are not limited to:

- System Hierarchy (functional decomposition)
- FFBD and Integrated Definition for Functional Modeling (IDEF) diagrams
- System Schematic
- N^2 Chart
- Operational Scenario and System ConOps documents
- Kepner-Tregoe Analysis (KTA) and Analytic Hierarchy Process (AHP) models (e.g. Expert Choice)
- Decision Trees.

The result of this effort is a set of System Architecture options that spans the region of design space containing the optimum, with sufficient descriptive and supporting documentation. Specific products include:

- For each System Architecture option, identification of the elements making up that option, their arrangement, the interactions among the elements, and a description of the salient features and parameter values, as necessary, to characterize the option for analysis and as a potential System Architecture baseline. This can take the form of diagrams, schematics, concept drawings, tabular data, and narrative. The ConOps should also be created for each candidate System Architecture (see Section 4.3.2.4).
- Documentation of the rationale which justifies the selection of the set of options.
- Documented assurance, with reasonable certainty for the set of candidate System Architecture options as a whole, that (1) the set spans the region of design space that contains the optimum, (2) the selected set will support efficient selection and closure on the optimum, and (3) the descriptive data (features and parameters) are adequate to support subsequent system definition work.

The following measures are used to gauge the progress and completion of the architecture synthesis activity:

1. Technical performance, schedule spans, costs, and risk estimates for each alternative
2. Evidence that each alternative is consistent with the business case for the system.

4.3.2.5 Analyze and Select Preferred System Architecture/Element Solution

The objective of this activity is to select or evolve the preferred System Architecture from the set of System Architecture options developed in the previous activities. The selection of the preferred System Architecture is essentially a tradeoff among the various architecture options, using the tradeoff process with modeling. It includes the possibility of combining the best features of several options and modifying top contenders to further improve their desirability. The selected baseline System Architecture should be robust (i.e., allows subsequent, more detailed system definition to proceed with minimum backtracking as additional information is uncovered) and acceptably close to the theoretical optimum in meeting requirements, with acceptable risk and within available resources.

1. Create models that map each option's characteristics onto measures of success against the criteria. – The models should be as objective and analytical as possible. However, to economize on the expended effort, the detail, depth of fidelity, and precision of the models need be sufficient only to clearly distinguish between the options (i.e., the models are used only to produce a clear ranking of the options and not as a design tool). Modeling, simulation, and prototyping are addressed in detail in Sections 4.3.2.6 and 4.12.1.

2. Use Trade Studies methods to compare and rank the options. – Frequently, a simple weighted scoring spreadsheet, with subjective evaluation of options against the criteria, will be adequate. With this method, the criteria are of two types: go/no-go criteria that must be met, and criteria used to evaluate the relative desirability of each option on a proportional scale. The go/no-go criteria are applied first as an initial screening. Any option that fails any of these criteria is ruled out of further consideration. Other useful tools include:
   a. KTA and AHP models (e.g. Expert Choice)
   b. Software for Multi-Attribute Utility Analysis (MAUA)
   c. Models for converting option parameters to scores against criteria.

3. Modify options or combine the best features of several options to correct shortcomings and advance the capability of the leading contenders. – Consider doing a Force Field Analysis and mitigation analysis to help in making decisions. Also, look for adverse consequences and potentially unacceptable risks associated with the top contenders, then either correct such conditions or eliminate options that cannot be corrected.

4. Perform sensitivity analysis to test the robustness of the final selection. – Examine the effects of variation in the definitions and application of the criteria, the methods of analyzing and evaluating the options, and any assumptions inherent in the analysis. Also look for plausible scenarios that could result in a different selection. If two or more of the options are closely ranked or the ranking can be changed by plausible means, then look for ways to arrive at a clear decision by strengthening the options or improving the selection method, perhaps by expanding the set of criteria.

5. Document the process. – Provide a clear description of how each step is implemented, justify all choices made, and state all assumptions.

The result of this effort is a System Architecture baseline, with sufficient descriptive and supporting documentation. Specific products include:

- The selected System Architecture baseline.
- Identification of the elements (type and principle design characteristics), their arrangement, the interactions among the elements, and a description of the system's salient features and parameter values, as necessary to characterize the System Architecture baseline. This can take the form of diagrams, schematics, concept drawings, operational and life-cycle scenarios, tabular data, and narrative.
- Documentation of the selection process to:
  - Justify the selection
  - Enable its review
  - Support subsequent system development, modification and growth throughout its life cycle.
- Documented assurance, with reasonable certainty, that the selected System Architecture baseline is adequately close to the theoretical optimum, that it is robust, and that the descriptive data (features and parameters) are adequate to support subsequent work.

The following measures are used to gauge the progress and completion of the preferred architecture selection activity:

1. Completeness of the selection criteria mentioned in paragraph 1 of the recommended activities
2. Completeness of the documentation.

4.3.2.6 Model, Simulate, and Prototype System Architectures

Modeling, simulation, and prototyping used during architecture design can significantly reduce the risk of failure in the finished system. These techniques enable the development of complex and costly enabling systems, such as a flight simulator or a high-volume production line, that allow validation of the system's concepts or supports training of personnel in ways that would otherwise be cost prohibitive. Systems engineers use modeling and simulation on large complex projects to manage the risk of failure to meet system mission and performance requirements. This form of analysis is best conducted by subject matter experts who develop and validate the models, conduct the simulations, and analyze the results. A detailed discussion of these cross-cutting techniques appears in Section 4.12.1.

4.3.2.7 Define, Refine, and Integrate System Physical Configuration

After the System Architecture has been selected, sufficient detail must be developed on the elements to (1) ensure that they will perform as an integrated system within their intended environment and (2) enable subsequent development or design activity, as necessary, to fully define each element. The objective is to allow the further definition of each CI to proceed on its own, in parallel with all the others. As such, systems engineers integrate the selected system architecture baseline and CIs, the customer's definition of external interfaces, and available technical data on interfacing items to establish the physical, software, and operational implementations at the next level of detail for all elements in the selected architecture. Additionally, interface parameters and, to the degree possible for the current stage of development, the values of those parameters are identified.

The recommended method is to establish a systematic framework for identifying interfaces and tracking descriptive data, acquiring updates as they occur, and displaying a consistent set of data in a uniform format to concerned parties. Useful tools for completing this process include N^2 charts, system schematics, interface diagrams, tables, and drawings of detailed interface data.

This activity can be led either by SE or by Design Engineering, depending on the technical maturity of the design at the time. In either case, the discipline not in the lead has a strong supporting role. Specific roles and responsibilities for this activity are as follows:

- *Systems Engineering* – provides the process for generating and selecting options for each CI, performs analyses and trades, identifies and coordinates interfaces, integrates the results, and ensures that all requirements are implemented.

- *Design Engineering* – creates design options for CIs and their arrangement as a system, develops technical definition data, performs analyses and trades, and documents design decisions.
- *Supporting Disciplines* – propose options for CIs or their features, monitor implementation of requirements in each specialty area, and review the results of the system definition process.

This activity often proceeds in parallel with the Define, Derive, and Refine Functional/Performance Requirements activity as system development and design proceeds into a more detailed level of definition. The steps for to define, refine, and integrate the system physical configuration are as follows:

1. Create a system-level description of system operation, using appropriate tools and notation, to enable a thorough analysis of the system's behavior at the interfaces among all of its elements. This includes preparing system interface diagrams for each interface.
2. Enter the available data about the elements into the system-level description. – Obtain interface identification and definition from design engineering and supporting disciplines. Determine what additional data are needed in order to support analysis of system operation.
3. Perform design activity on the elements as needed to provide the additional data needed for the system-level description (see Task 2).
4. Perform liaison with customer representatives regarding definition of interfaces with the system's operating environment throughout its life cycle.
5. Analyze system operation to verify its compliance with requirements. Modify elements and system architecture, and resolve interface issues as needed to bring the result into compliance.

The result of this activity is a selected set of design concepts for CIs (including selected technologies, configurations, design parameter values, and arrangements) to implement all of the system elements as an integrated system. This includes documented definitions of all system interfaces and documented justification for the selected concepts.

The following measures are used to gauge the progress and completion of the preferred architecture selection activity:

1. System requirements not met (if any) by selected concept.
2. Number or percent of system requirements verified by system operation analyses.

3. Number of TBD/TBR requirements in system architecture or design.
4. Number of interface issues not resolved.
5. Percent of identified system elements that have been defined.

4.3.2.8 Implement Requirements and Design Feedback Loops

Design is the process of defining, selecting, and describing solutions to requirements in terms of products and processes. A design describes the solution (conceptual, preliminary, or detailed) to the requirements of the system. Synthesis is the translation of input requirements (including performance, function, and interface) into possible solutions satisfying those inputs. Synthesis defines a physical architecture of people, product, and process solutions for logical groupings of requirements and then designs architectures for those solutions. This section describes the iterations and feedback ("loops") between the requirements and design activities. The loop is also an integrated process to refine the requirements.

The key participants in carrying out the requirements and design feedback loops are SE, with support from design, manufacturing, specialty engineering, and materials and processes engineering. The role of SE is to ensure that the proper inputs and feedback to hardware and software are occurring at the system and lower levels. Inputs include the outputs from the previous stage, the results of Requirements Analysis Process, the project baseline, and any proposed changes (initiated by customer or internally from requirements and design analyses, new technology or test results).

The result of these feedback loops is system and lower-level designs that are properly allocated to hardware and software and thoroughly audited to ensure that they meet requirements and are consistent with established manufacturing practices.

Systems engineering activities conducted in implementing Requirements and Design Feedback Loops are as follows:

1. Determine how the SE process is tailored (see Chapter 8) for different levels of the project. – This SE task is performed in conjunction with project management and determines the amount and detail of SE to be performed at each level. This should be established early in the project and is covered in the Systems Engineering Plan (SEP; see Section 5.1.2.2).
2. Audit the system requirements. – Audits occur at various levels, from peer reviews against requirements in specifications to design reviews, both informal and formal. The results of the audits serve as feedback to previous SE activities may cause changes in requirements at any

level. The output of any process stage becomes input to the next stage and should include an audit trail of requirements, designs, and decisions.

3. Conduct design reviews at the appropriate points in the development effort. – Some of these design reviews can be part of formal decision gates. Results of these design reviews should be flowed back into the requirements develop process.

4. Iterate between systems (i.e., hardware and software), design, and manufacturing functions. – SE should ensure that Integrated Product and Process Development (IPPD) engineering is taking place. This may be by chairing an IPDT, or by being a member of one. In either case, it is the responsibility of SE to ensure that all necessary disciplines in the project are participating in any stage. SE consults on all stages of the project to provide the traceability and flowdown of the customer's needs and requirements. As necessary, SE will conduct producibility meetings and producibility trade studies to determine production methods and materials.

5. Audit the design and manufacturing process. – After the build-to decision gate, SE audits the design (hardware and software) and manufacturing processes to ensure compliance with requirements. Audits occur at various levels, from peer reviews to design reviews, both informal and formal. This provides feedback to the requirements and design functions.

6. Iterate with other parts of the SE process. – As indicated above, SE ensures that all the elements of the SE process are executed.

7. Interface with specialty engineering groups and subcontractors to ensure common understanding across disciplines. – This is part of the SE role in ensuring that IPPD engineering is being performed on the project.

8. Update models as better data becomes available. – SE should always ensure that models are up to date.

Standard Configuration Management processes document a baseline that is consistent with the output of the project. Alternatively, SE should create a baseline document that contains drawings, specifications, published analyses, and deliverable documents to show the current baseline. Care should be taken to ensure that all internal and external interfaces and interactions are included. Proposed and actual changes to the project baseline can also prompt a feedback loop to be invoked.

The results of this activity include a new project baseline consisting of requirements, specifications, and a producible design that has been audited to ensure compliance with requirements.

The following criteria indicate completion of the requirements and design feedback activity:

- Successfully establish project baseline
- Completion of requirements audits
- Completion of design audits.

4.4 Implementation Process

4.4.1 Overview

4.4.1.1 Purpose

As stated in ISO/IEC 15288:2008:

> *The purpose of the Implementation Process is to realize a specified system element.*
>
> *This process transforms specified behaviour, interfaces and implementation constraints into fabrication actions that create a system element according to the practices of the selected implementation technology. The system element is constructed or adapted by processing the materials and/or information appropriate to the selected implementation technology and by employing appropriate technical specialties or disciplines. This process results in a system element that satisfies specified design requirements through verification and stakeholder requirements through validation.*[25]

In short, the Implementation Process designs, creates, or fabricates a system element conforming to that element's detailed description. The element is constructed employing appropriate technology and industry practices.

4.4.1.2 Description

During the Implementation Process, engineers follow the requirements allocated to the system element to design, fabricate, code, or build each individual element using specified materials, processes, physical or logical arrangements, standards, technologies, and/or information flows outlined in detailed drawings or other design documentation. Requirements are verified and stakeholder requirements are validated. If subsequent configuration audits reveal discrepancies, recursive interactions occur with predecessor activities or processes, as required, to correct them. Figure 4-12 is the context diagram for the Implementation Process.

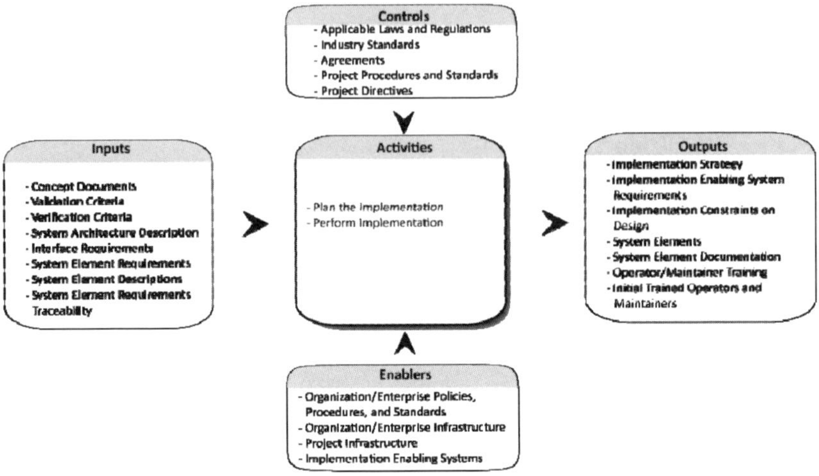

Figure 4-12 Context Diagram for the Implementation Process

4.4.1.3 Inputs

The inputs to the Implementation Process include the following:

- The primary inputs to the Implementation Process are the system architecture and element baseline documented during the Architectural Design Process and other information generated in the Requirements Analysis and Stakeholder Requirements Definition Processes, including:
  - *Concept Documents*
  - *Validation Criteria*
  - *Verification Criteria*
  - *System Architecture*
  - *Interface Requirements*
  - *System Element Requirements*
  - *System Element Descriptions*
  - *System Element Requirements Traceability.*

This process is governed by the following controls and enablers:

- *Applicable Laws and Regulations*

- *Industry Standards* – relevant industry specifications and standards
- *Agreements* – terms and conditions of the agreements, especially those related to packaging, handling, storage, and transportation (PHS&T) and initial operator training
- *Project Procedures and Standards* – including project plans
- *Project Directives*
- *Organization/Enterprise Policies, Procedures, and Standards* – including guidelines and reporting mechanisms including safety practices and other guidelines
- *Organization/Enterprise Infrastructure*
- *Project Infrastructure*
- *Implementation Enabling Systems* – examples include unique facilities for hardware fabrication and software development.

4.4.1.4 Outputs

Outputs of the Implementation Process include the following:

- *Implementation Strategy* – May also include inputs to determine integration constraints
- *Implementation Enabling System Requirements* – Requirements for any systems needed to enable implementation of the system-of-interest need to be developed
- *Implementation Constraints on Design* – Any constraints on the design arising from the implementation strategy
- *System Elements* – Verified, validated, and supplied according to agreement
- *System Element Documentation*, including:
  - Detailed drawings, codes, and material specifications
  - Updated design documentation, as required by corrective action or adaptations caused by acquisition or conformance to regulations
- *Operator/Maintainer Training* – Training capabilities and documentation
- *Initial Trained Operators and Maintainers* – Staff of trained operators, maintainers, and support personnel, according to the agreement.

4.4.1.5 Process Activities

Implementation Process activities begin with detailed design and include the following:

- *Plan the Implementation*
  - Develop an Implementation Strategy – Define fabrication/coding procedures, tools and equipment to be used, implementation tolerances, and the means and criteria for auditing configuration of resulting elements to the detailed design documentation. In the case of repeated system element implementations (such as for mass manufacturing or replacement elements), the implementation strategy is defined/refined to achieve consistent and repeatable element production and retained in the project decision database for future use.
- *Perform Implementation*
  - Develop data for training users on correct and safe procedures for operating and maintaining that element — Either as a stand-alone end item or as part of a larger system.
  - Complete detailed product, process, material specifications ("Build-to" or "Code-to" documents) and corresponding analyses and produce documented evidence of Implementation compliance — Specifically, these tasks are as follows:
    - Conduct peer reviews and testing – Inspect and verify software for correct functionality, white box testing, etc. in accordance with software/hardware best practices.
    - Conduct hardware conformation audits – Compare hardware elements to detailed drawings to ensure that each element meets its detailed specifications prior to integration with other elements in higher CIs or assemblies.
  - Prepare initial training capability and draft training documentation - to be used to provide the user community with the ability to operate, conduct failure detection and isolation, and maintain the system as appropriate.
  - Prepare a hazardous materials log, if applicable.
  - Train initial operators and maintainers on the use of elements that provide a human-system interface or require maintenance actions at the element level.

Common approaches and tips:

- Keep the IPDT engaged to assist with configuration issues and redesign.
- Inspections are a proactive way to build in quality.[26]
- In anticipation of improving process control, reducing production inspections, and lowering maintenance activities, many manufacturing firms use Design for Six Sigma or Lean Manufacturing.
- Conduct hardware conformation audits or system element level hardware verification; ensure sufficient software unit verification prior to integration.
- Validate simulations; interface simulator drivers should be representative of tactical environments.

4.4.2 Elaboration

4.4.2.1 Implementation Concepts

The Implementation Process typically focuses on the following three forms of system elements:

- *Hardware/Physical* – Output is fabricated hardware or physical element
- *Software* – Output is software code and executable images
- *Humans (Operators & Maintainers)* – Output is procedures and training.

The Implementation Process can support either the adaptation or realization of system elements. For system elements that are reused or acquired, such as COTS, the Implementation Process allows for adaption of the elements to satisfy the needs of the system-of-interest. This is usually accomplished via configuration settings provided with the element (e.g., hardware configuration switches and software configuration tables). Realized products have more flexibility to be designed and developed to meet the needs of the system-of-interest without modification.

4.5 Integration Process

4.5.1 Overview

4.5.1.1 Purpose

As stated in ISO/IEC 15288:2008:

The purpose of the Integration Process is to assemble a system that is consistent with the architectural design.

This process combines system elements to form complete or partial system configurations in order to create a product specified in the system requirements.[27]

This process is iterated with the V&V Processes, as appropriate.

4.5.1.2 Description

The Integration Process includes activities to perform the integration of system elements (hardware/physical, software, and procedures) and the demonstration of end-to-end operation (system build). System build is bottom-up. That is, elements at the bottom of the system hierarchy are integrated and verified first. This process verifies that all boundaries between system elements have been correctly identified and described, including physical, logical, and human-system interfaces and interactions (physical, sensory, and cognitive), and that all system element functional, performance, and design requirements and constraints are satisfied. Interim assembly configurations are verified to ensure correct flow of information and data across internal and external interfaces to reduce risk and minimize errors and time spent isolating and correcting them. Figure 4-13 is the context diagram for the Integration Process.

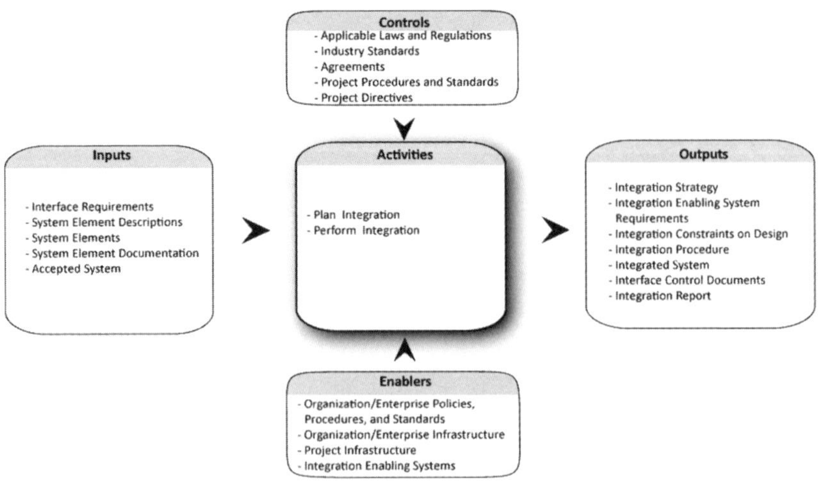

Figure 4-13 Context Diagram for the Integration Process

4.5.1.3 Inputs

Inputs to the Integration Process include the following:

- *Interface Requirements* – Includes applicable internal and external system element interface specifications
- *System Elements* – Includes the following associated information:
  - *System Element Descriptions*
  - *System Element Documentation*
- *Accepted System* – If the system-of-interest includes acquired system elements.

This process is governed by the following controls and enablers:

- *Applicable Laws and Regulations*
- *Industry Standards* – relevant industry specifications and standards
- *Agreements* – terms and conditions of the agreements
- *Project Procedures and Standards* – including project plans
- *Project Directives*
- *Organization/Enterprise Policies, Procedures, and Standards* – including guidelines and reporting mechanisms
- *Organization/Enterprise Infrastructure*
- *Project Infrastructure*
- *Integration Enabling Systems* – examples include integration tools, facilities, and verification equipment, including any integration technology constraints.

4.5.1.4 Outputs

Outputs of the Integration Process include the following:

- *Integration Strategy*
- *Integration Enabling System Requirements* – Requirements for any systems needed to enable integration of the system-of-interest need to be developed
- *Integration Constraints on Design* – Any constraints on the design arising from the integration strategy
- *Integration Procedure*
- *Integrated System* – Completed subsystem or system ready for verification
- *Interface Control Documents (ICDs)*

- *Integration Report* – Including documentation of the Integration testing and analysis results, areas of non-conformance, and validated internal interfaces.

The Integration Process may also support the updating of affected documentation from other processes, including:

- Updated Interface Specifications
- Updated RVTM
- Updated product assembly drawings
- Updated manufacturing tool drawings.

4.5.1.5 Process Activities:

The Integration Process includes the following activities:

- *Plan Integration*
  - Define the integration strategy
  - Schedule integration testing tools and facilities
- *Perform Integration*
  - Assemble system elements according to the integration plan
  - Validate and Verify Interfaces – Confirm correct flow of information across internal interfaces through "black box testing" at each successive level of assembly
  - Verify and analyze assemblies – Confirm correct functionality of assembled products through integration testing and analysis at each successive level of assembly
  - Document integration testing and analysis results
  - Document and control the architectural baseline – Includes capturing any modifications required during this process.

Common approaches and tips:

- Keep the IPDT engaged to assist with configuration issues and redesign.
- Maintain configuration control over drawings, specifications, interface control drawings, and published analyses.
- Define an integration strategy that accounts for the schedule of availability of system elements (including the humans that will use,

operate, maintain, and sustain the system) and is consistent with fault isolation and diagnosis engineering practices.

4.5.2 Elaboration

4.5.2.1 Integration Concepts

The System Integration activity establishes system internal interfaces and interfaces between the system and larger program(s). The Systems Integration activity includes the integration and assembly of the system with emphasis on risk management and continuing verification of all external and internal interfaces (physical, functional, and logical).

At the top level, System Integration is performed on the system and its elements, and on the system and interfacing external systems. The objective is to ensure that elements are integrated into the system and that the system is fully integrated into the larger program. The System Integration activity may also be involved in the integration of subsystems into their elements. At lower levels of the system hierarchy, Product Integration Teams (PITs) and Product Development Teams (PDTs) perform integration.

A discussion of these activities is divided to address the internal interfaces among the elements comprising the system (i.e., System Build) and the external interfaces between the system and other systems (i.e., System Integration with External Systems).

4.5.2.2 System Build

This process addresses the integration internal to the system (i.e., the integration of all the elements comprising the system). System build is bottom-up. That is, elements at the bottom of the system hierarchy are integrated and verified first. Tasks associated with the System Build activity are as follows:

1. Obtain the system hierarchy – The system hierarchy shows the relationship between the system segments and elements, which are structured functionally to form the system. The activity begins with a good knowledge of this system structure. In addition to the system hierarchy, obtain the system and CI design specifications, functional block diagrams, N^2 charts, and any other data that defines the system structure and its interfaces.

2. Determine the interfacing system elements.

3. Ascertain the functional and physical interfaces of the system and system elements – This will require a detailed assessment of the functions flowing in both directions across the interfaces, such as data, commands, and power. It will also require a detailed assessment of

the physical interfaces, such as fluids, heat, mechanical attachments and footprints, connectors, and loads.

4. Organize ICDs or drawing(s) to document the interfaces and to provide a basis for negotiating the interfaces between the parties to the interfaces.

5. Work with producibility/manufacturing groups to verify functional and physical internal interfaces and to ensure changes are incorporated into the specifications.

6. Conduct internal IFWGs, as required. – These groups involve all the relevant engineering disciplines. There may be a series of subgroups by discipline, or one group, depending on the size and complexity of the system.

7. Review test procedures and plans that verify the interfaces.

8. Audit design interfaces.

9. Ensure that interface changes are incorporated into specifications.

4.5.2.3 System Integration with External Systems

This activity addresses the System Integration external to the system (i.e., the integration of all the system under development with interfacing external systems). Standard Configuration Management Processes document a baseline that is consistent with the output of the project. Alternatively, SE should create a baseline document that contains drawings, specifications, published analyses, and deliverable documents to show the current baseline. Care should be taken to ensure that all internal and external interfaces and interactions are included.

IFWGs are established to review interface statements/drawings, and are a good means of ensuring direct interaction of all parties to the interface, as discussed above. The following tasks are conducted to integrate the system-of-interest with external systems:

1. Obtain the system hierarchy, the systems and CI design specifications, functional block diagrams, N^2 charts, and any other data that define the system structure and its interfaces.

2. Determine the interfacing systems by reviewing the items in step 1 above.

3. Obtain interfacing programs' ICDs, SEPs, and other relevant interface documents. The ICD is developed over a series of meetings/teleconferences in which the representatives of each side of the interface directly present the performance or needs for their side of the interface. One party takes the lead to be the author of the ICD

and to ensure that copies are available to other parties before a meeting. All parties sign the ICD when agreement has been reached, after which it is released and comes under formal change control.

4. Ascertain the functional and physical interfaces of the external systems with the subject system. This will require a detailed assessment of the functions flowing in both directions across the interface, such as data, commands, and power. It will also require a detailed assessment of the physical interfaces, such as fluids, heat, mechanical attachments and footprints, connectors, and loads.
5. Organize an ICD to document the interfaces and to provide a basis for negotiating the interfaces between the parties to the interfaces.
6. Conduct IFWGs among the parties to the interfaces. These can be one group covering all interfaces for a smaller program, or it can be broken into engineering disciplines addressing the interfaces for larger programs.
7. Review test procedures and plans that verify the interfaces.
8. Audit design interfaces.
9. Incorporate interface changes into specifications.

The following measures are used to gauge the progress and completion of the systems integration task:

- Percentage of released interface drawings
- Percentage of completed ICDs
- Percentage of approved ICDs
- Number and type of interface issues resolved and unresolved.

4.6 Verification Process

4.6.1 Overview

4.6.1.1 Purpose

As stated in ISO/IEC 15288:2008:

> *The purpose of the Verification Process is to confirm that the specified design requirements are fulfilled by the system.*
>
> *This process provides the information required to effect the remedial actions that correct non-conformances in the realized system or the processes that act on it.*[28]

4.6.1.2 Content/Description

The Verification Process confirms that the system-of-interest and all its elements perform their intended functions and meet the performance requirements allocated to them (i.e., that the system has been built right). Verification methods include inspection, analysis, demonstration, and verification and are discussed in more detail below. Verification activities are determined by the perceived risks, safety, and criticality of the element under consideration.

The Verification Process works closely with other life cycle processes. A key outcome of the Planning Process is the creation of project procedures and processes that specify the forms of system assessments (e.g., conformation audits, integration, verification, and validation) in appropriate project documents (e.g., SEPs, schedules, and specifications). Specification of verification criteria takes place as the requirements are written, but the creation of a procedure to assess compliance is part of this process. Figure 4-14 is the context diagram for the Verification Process.

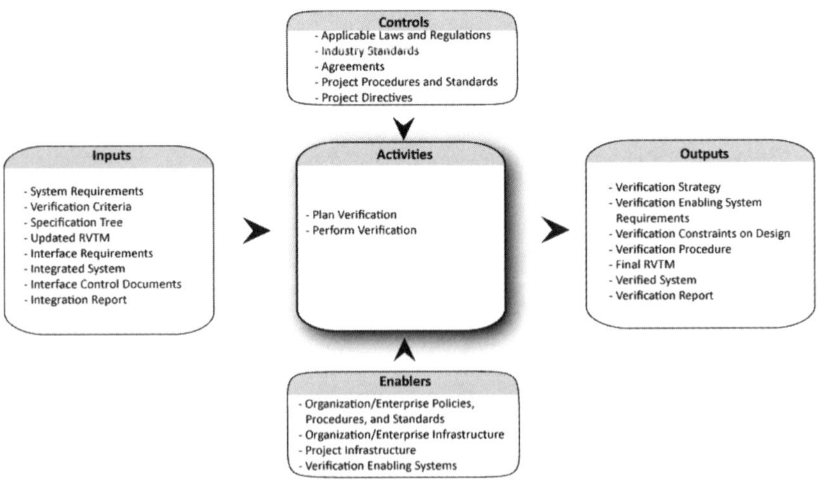

Figure 4-14 Context Diagram for the Verification Process

4.6.1.3 Inputs

Primary inputs to the Verification Process include the following:

- Project baseline documented during the Requirements Analysis and Architectural Design Processes, including:
    - *System Requirements*
    - *Verification Criteria*

   - *Specification Tree*
   - *Updated RVTM*
   - *Interface Requirements*
- Project baseline documented during the Integration Processes, including:
   - *Integrated System*
   - *ICDs*
   - *Integration Report.*

This process is governed by the following controls and enablers:

- *Applicable Laws and Regulations*
- *Industry Standards* – relevant industry specifications and standards
- *Agreements* – terms and conditions of the agreements
- *Project Procedures and Standards* – including project plans
- *Project Directives*
- *Organization/Enterprise Policies, Procedures, and Standards* – including guidelines and reporting mechanisms
- *Organization/Enterprise Infrastructure*
- *Project Infrastructure*
- *Verification Enabling Systems* – examples include test facilities and test equipment.

4.6.1.4 Outputs

Outputs of the Verification Process include the following:

- *Verification Strategy*
- *Verification Enabling System Requirements* – The requirements for any systems needed to enable verification of the system-of-interest need to be developed
- *Verification Constraints on Design* – Identify any constraints on the design arising from the verification strategy
- *Verification Procedure*
- *Final RVTM* – The RVTM is updated accordingly, and eventually finalized, as independent verification actions are performed

- *Verified System* – Completed subsystem or system ready for verification
- *Verification Report* – Including documentation of the verification results, a record of any recommended corrective actions, Design Feedback/Corrective Actions taken, and evidence that the system element or system satisfies the requirements, or not.

4.6.1.5 Process Activities

The Verification Process includes the following activities:

- *Plan Verification*
  - Schedule, confirm, and install verification enabling systems
- *Perform Verification*
  - Develop verification procedures
  - Conduct verification activities, per established procedures, to demonstrate compliance with requirements

    Document verification results and enter data into the RVTM.

Common approaches and tips:

- The RVTM is frequently used as a single point of accountability for tracing a requirement back to the source of the need and forward through the life cycle to assess that the need has been met.
- Beware the temptation to reduce verification activities due to budget or schedule overruns. – Remember the message of Figure 2-4, namely, that discrepancies and errors are more costly to correct later in the system life cycle.[29]
- Avoid conducting verification late in the schedule when there is less time to handle discrepancies, or too early, before development is complete.

4.6.2 Elaboration

4.6.2.1 Verification Concepts

System verification ensures that the system, its elements, and its interfaces conform to their requirements; in other words that “you built it right.” Verification encompasses the tasks, actions, and activities performed to evaluate the progress and effectiveness of the evolving system solutions (people, products, and process) and to measure compliance with requirements. The primary purpose of verification is to determine that system specifications, designs, processes, and products are compliant with requirements. A continuous feedback of verification data helps to reduce risk and to surface problems early. The goal is to completely verify system capability to meet all requirements prior to production and operation stages. Problems uncovered at in these stages are very costly to correct (see Figure 2-4). As such, early discovery of deviations from requirements reduces overall project risk and helps the project deliver a successful, low-cost system.[29] Verification results are an important element of decision gate reviews.

Verification analysis can be initiated once a design concept has been established. If an RVTM is used, each requirement has a verification activity associated with it. A unique requirements identifier can be used for traceability to the verification plans, verification procedures, and verification reports to provide a closed loop Verification Process from demonstrated capability back to the requirement. Basic verification activities are as follows:

- *Inspection* – An examination of the item against applicable documentation to confirm compliance with requirements. Inspection is used to verify properties best determined by examination and observation (e.g., paint color, weight, etc.).
- *Analysis* – Use of analytical data or simulations under defined conditions to show theoretical compliance. Analysis (including simulation) is used where verifying to realistic conditions cannot be achieved or is not cost-effective and when such means establish that the appropriate requirement, specification, or derived requirement is met by the proposed solution.
- *Demonstration* – A qualitative exhibition of functional performance, usually accomplished with no or minimal instrumentation. Demonstration (a set of verification activities with system stimuli selected by the system developer) may be used to show that system or subsystem response to stimuli is suitable (see Figure 4-15). Demonstration may also be appropriate when requirements or specifications are given in statistical terms (e.g., mean time to repair, average power consumption, etc.).
- *Test* – An action by which the operability, supportability, or performance capability of an item is verified when subjected to controlled conditions that are real or simulated. These verifications often use special test equipment or instrumentation to obtain very accurate quantitative data for analysis.
- *Certification* - Written assurance that the product or article has been developed and can perform its assigned functions in accordance with

legal or industrial standards. The development reviews and verification results form the basis for certification; however, certification is typically performed by outside authorities, without direction as to how the requirements are to be verified. For example, this method is used for electronics devices via CE certification in Europe and UL certification in the United States and Canada.

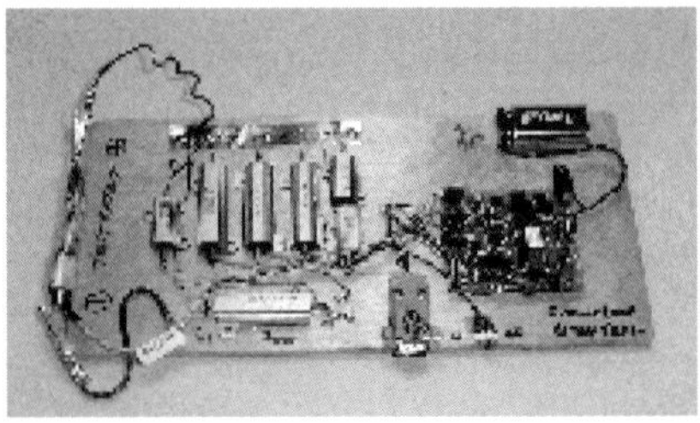

Figure 4-15 Test platform for analyzing battery performance at high loads[31]

Verification can also be done by similarity in certain situations. Similarity is most appropriate where a design is being modified or is very similar to an existing verified system. When verifying by similarity, a common scenario is to perform an analysis to ensure the design features and operational environment of the system-of-interest are similar enough to not warrant the expense of the verification demonstration or test.

The design of the verification activity involves choosing the most cost-effective mix of simulations and physical and integrating test results to avoid unnecessary redundancy. Complete simulation of the system (both performance and design) has become common-place in major system development and has resulted in reduced development time and cost.

There are four basic test categories. They are:

- *Development Test* – Conducted on new items to demonstrate proof of concept or feasibility.
- *Qualification Test* – Conducted to prove that the system design meets its requirements with a predetermined margin above expected operating conditions, for instance by using elevated environmental conditions for hardware.
- *Acceptance Test* – Conducted prior to transition such that the customer can decide that the system is ready to change ownership from supplier to acquirer.

- *Operational Test* – Conducted to verify that the item meets its specification requirements when subjected to the actual operational environment.

Verification analysis checks that tests have been established using realistic scenarios to demonstrate human reaction times that satisfy operational requirements. Maintainability demonstrations should include a sufficient number of tests and problem areas to provide a high confidence level of meeting maintainability parameters, such as Mean-Time-To-Repair. Production line tests are recommended for items that are new or have not been previously applied to this application. The tests demonstrate producibility and repeatability. Human Factors Engineers develop task descriptions and operational sequence diagrams and evaluate the human-system interface to establish the required interactions with the hardware and software. These products can be leveraged to support verification analysis activities.

4.7 Transition Process

4.7.1 Overview

4.7.1.1 Purpose

As stated in ISO/IEC 15288:2008:

> *The purpose of the Transition Process is to establish a capability to provide services specified by stakeholder requirements in the operational environment.*
>
> *This process installs a verified system, together with relevant enabling systems, e.g., operating system, support system, operator training system, user training system, as defined in agreements. This process is used at each level in the system structure and in each stage to complete the criteria established for exiting the stage. It includes preparing applicable storage, handling, and shipping enabling systems.*[32]

Ultimately, the Transition Process transfers custody of the system and responsibility for system support from one organizational entity to another. This includes, but is not limited to, transfer of custody from the development team to the organizations that will subsequently operate and support the system. Successful conclusion of the Transition Process typically marks the beginning of the Utilization Stage of the system-of-interest.

4.7.1.2 Description

The Transition Process installs a verified system in the operational environment along with relevant enabling systems, such as operator training systems, as defined in the agreement. As part of this process, the acquirer accepts that the system provides the specified capabilities in the intended operational environment prior to allowing a change in control, ownership, and/or custody. While this is a relatively short process, it should be carefully planned to avoid surprises and recrimination on either side of the agreement. Additionally, transition plans should be tracked and monitored to ensure all activities are completed to both parties' satisfaction, including resolution of any issues arising during transition. Figure 4-16 is the context diagram for the Transition Process.

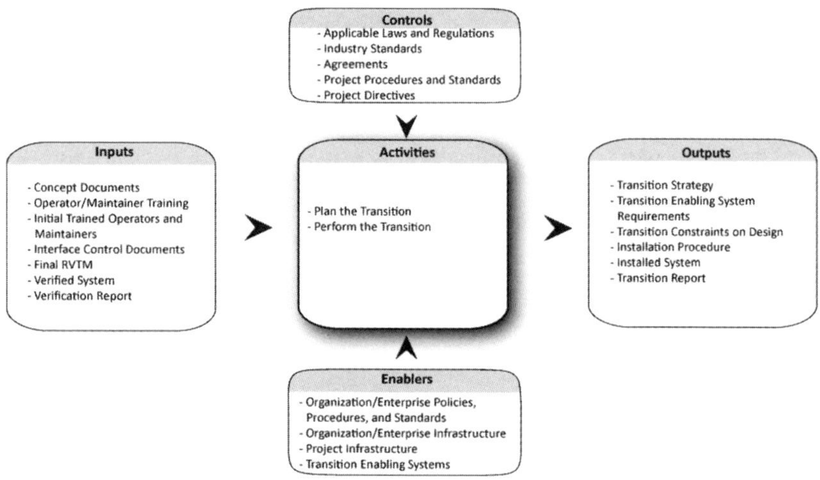

Figure 4-16 Context Diagram for the Transition Process

4.7.1.3 Inputs

Inputs to the Transition Process include the following:

- *Concept Documents*
- *Operator/Maintainer Training*
- *Initial Trained Operators and Maintainers* – Commissioning of the system includes the humans that will operate, maintain, and sustain the system.
- *ICDs*
- *Final RVTM*
- *Verified System* – Availability of the system-of-interest together with enabling systems is prerequisite to beginning the Transition Process.
- *Verification Report.*

This process is governed by the following controls and enablers:

- *Applicable Laws and Regulations*
- *Industry Standards* – relevant industry specifications and standards
- *Agreements* – terms and conditions of the agreements
- *Project Procedures and Standards* – including project plans
- *Project Directives*
- *Organization/Enterprise Policies, Procedures, and Standards* – including guidelines and reporting mechanisms
- *Organization/Enterprise Infrastructure*
- *Project Infrastructure*
- *Transition Enabling Systems* – examples include storage, handling, shipping, and training systems.

4.7.1.4 Outputs

Outputs of the Transition Process include the following:

- *Transition Strategy*
- *Transition Enabling System Requirements* – Requirements for any systems needed to enable transition of the system-of-interest need to be developed
- *Transition Constraints on Design* – Any constraints on the design arising from the transition strategy
- *Installation Procedure*
- *Installed System* –The system is installed and acceptance criteria are met or discrepancies documented with recommended and agreed upon corrective actions.
- *Transition Report* – Including documentation of the transition results and a record of any recommended corrective actions, such as limitations, concessions, and on-going issues. The transition report should also include plans to rectify any problems that arise during transition.

4.7.1.5 Process Activities

The Transition Process includes the following activities:

- *Plan the Transition*

  - Prepare a transition strategy, including operator training, logistics support, delivery strategy, and problem rectification/ resolution strategy.
  - Develop installations procedures
- *Perform the Transition*
  - Prepare the installation site and Install system per established procedures
  - Train the users in the proper use of the system and affirm users have the knowledge and skill levels necessary to perform Operation and Maintenance activities. This includes a complete review and handoff of operator and maintenance manuals, as applicable.
  - Receive final confirmation that the system—as operated and maintained by the intended users—meets their needs. This process typically ends with a formal, written acknowledgement that the system has been properly installed and verified, that all issues and action items have been resolved, and that all agreements pertaining to development and delivery of a fully supportable system have been fully satisfied or adjudicated.
  - Post-implementation problems are documented and may lead to corrective actions or changes to the requirements.

Common approaches and tips:

- When acceptance activities cannot be conducted within the operational environment, a representative locale is selected.
- This process relies heavily on quality assurance (QA) and configuration management documentation.

4.7.2 Elaboration

Elaborations for this section are under development and will be included in a future revision of the *INCOSE Systems Engineering Handbook*.

4.8 Validation Process

4.8.1 Overview

4.8.1.1 Purpose

As stated in ISO/IEC 15288:2008:

The purpose of the Validation Process is to provide objective evidence that the services provided by a system when in use comply with stakeholders' requirements, achieving its intended use in its intended operational environment.

This process performs a comparative assessment and confirms that the stakeholders' requirements are correctly defined. Where variances are identified, these are recorded and guide corrective actions. System validation is ratified by stakeholders.[33]

This process is invoked during the Stakeholders Requirements Definition Process to confirm that the requirements properly reflect the stakeholder needs and to establish validation criteria (i.e., that the right system has been built). This process is also invoked during the Transition Process to handle the acceptance activities.

4.8.1.2 Description

In-process validation starts with a comparative assessment as a means to determine if stakeholders' requirements and defined MOEs have been correctly translated into technical design specifications and MOPs. Validation criteria are selected based on the perceived risks, safety, and criticality. Figure 4-17 is the context diagram for the Validation Process.

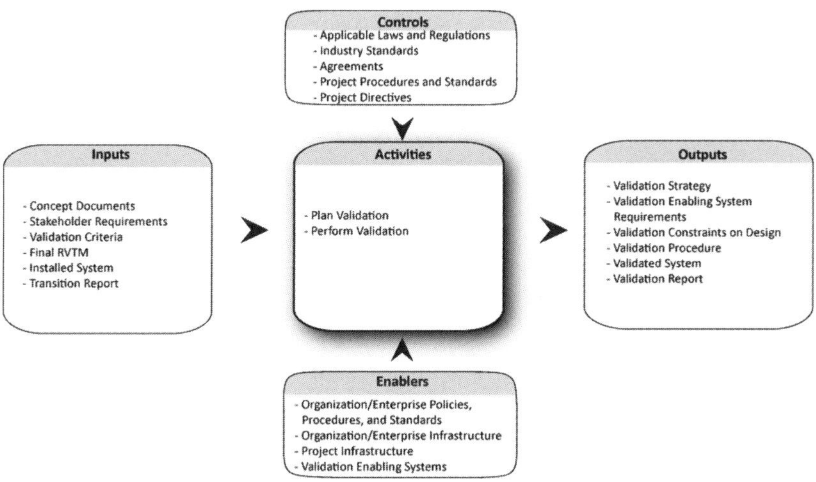

Figure 4-17 Context Diagram for the Validation Process

4.8.1.3 Inputs

Primary inputs to the Validation Process include the following:

- The project baseline documented during the Stakeholder Requirements Definition Processes:
  - *Concept Documents*
  - *Stakeholder Requirements* – When stakeholder requirements are elicited, validation criteria are applied before proceeding.
  - *Validation Criteria* – After the system-of-interest is verified, it is subjected to the validation criteria.
- The project baseline documented during the Verification and Transition Processes:
  - *Final RVTM*
  - *Installed System*
  - *Transition Report.*

This process is governed by the following controls and enablers:

- *Applicable Laws and Regulations*
- *Industry Standards* – relevant industry specifications and standards
- *Agreements* – terms and conditions of the agreements
- *Project Procedures and Standards* – including project plans
- *Project Directives*
- *Organization/Enterprise Policies, Procedures, and Standards* – including guidelines and reporting mechanisms
- *Organization/Enterprise Infrastructure*
- *Project Infrastructure*
- *Validation Enabling Systems* – examples include test facilities and test equipment.

4.8.1.4 Outputs

Outputs of the Validation Process include the following:

- *Validation Strategy*
- *Validation Enabling System Requirements* – Requirements for any systems needed to enable validation of the system-of-interest need to be developed
- *Validation Constraints on Design* – Any constraints on the design arising from the validation strategy

- *Validation Procedure*
- *Validated System* – Approved system baseline
- *Validation Report* – Including documentation of the validation activity results, a record of any recommended corrective actions, Design Feedback/Corrective Actions taken, and evidence that the system element or system satisfies the requirements, or not.

4.8.1.5 Process Activities

The Validation Process includes the following activities:

- *Plan Validation*
  - Develop a validation strategy
- *Perform Validation*
  - Develop validation procedures that demonstrate that the system is fit for its purpose and satisfies the stakeholders' requirements
  - Ensure readiness to conduct validation – system, enabling systems, and trained operators
  - Support in-process validation throughout system development
  - Conduct validation to demonstrate conformance to stakeholder requirements
  - If anomalies are detected, analyze for corrective actions and detect trends in failure to find threats to the system and evidence of design errors
  - Recommend corrective actions and obtain stakeholder acceptance of validation results
  - Document validation results and enter data into the RVTM.

Common approaches and tips:

- Validation methods during the Concept Stage include developing assessment scenarios that exercise all system modes and demonstrating system-level performance over the entire operating regime. The system design team uses the results of this activity to forecast success in meeting the expectations of users and the acquirer, as well as to provide feedback to identify and correct performance deficiencies before implementation.[29]

4.8.2 Elaboration

4.8.2.1 Validation Concepts

System validation confirms that the system, as built (or as it will be built), satisfies the stakeholders' stated needs. Validation ensures the requirements and the system implementation provide the right solution to the customer's problem. In other words, "you built the right thing" (verification, on the other hand, means that "you built the thing right" [see Section 4.6]).

Validation determines that a system does all the things it should and does not do what it should not do. End users and other stakeholders are usually involved in validation activities, but when warranted, an independent third party may be called in to perform validation. Validation may take place either in the operational environment or a simulated operational environment if conditions are hazardous. V&V activities often run concurrently and may use different portions of the same environment.

Requirements validation is conducted as part of requirements elicitation to provide early assurance that the requirements are the "right" requirements for guiding the development process to a conclusion that satisfies the stakeholders. Requirements validation is often based on requirements analysis; exploration of requirements adequacy and completeness; assessment of prototypes, simulations, models, scenarios, and mock-ups; and by obtaining feedback from customers, users, or other stakeholders. Much of the discussion regarding verification (see Section 4.6) can also be applied to validation.

The objects of validation are the designs, prototypes, and final systems elements, as well as the documentation and training materials that describe the system and how to use it. Validation results are an important element of decision gate reviews.

4.9 Operation Process

4.9.1 Overview

4.9.1.1 Purpose

As stated in ISO/IEC 15288:2008:

> *The purpose of the Operation Process is to use the system in order to deliver its services.*
>
> *This process assigns personnel to operate the system, and monitors the services and operator-system performance. In order to sustain services it identifies and analyzes operational problems in relation to agreements, stakeholder requirements and organizational constraints.*[34]

This process is often executed concurrent with the Maintenance Process.

4.9.1.2 Description

The Operation Process sustains system services by supplying personnel to operate the system, monitoring operator-system performance, and monitoring the system performance. When the system replaces an existing system, it may be necessary to manage the migration between systems such that persistent stakeholders do not experience a breakdown in services.

The Utilization and Support Stages of a system usually accounts for the largest portion of the total LCC. If system performance falls outside acceptable parameters, this may indicate the need for corrective actions in accordance with the Concept of Support and any associated agreements. When the system or any of its constituent elements reach the end of their planned or useful life, the system may enter the Disposal Process (see Section 4.11). Figure 4-18 is the context diagram for the Operation Process.

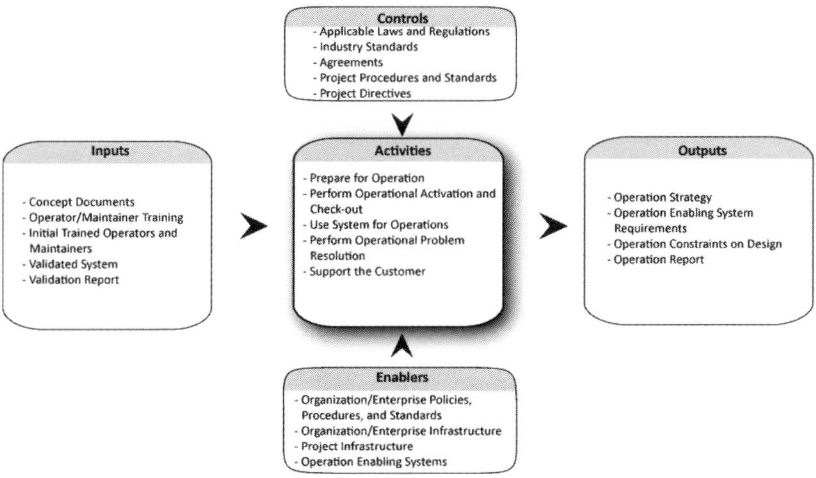

Figure 4-18 Context Diagram for the Operation Process

4.9.1.3 Inputs

Inputs to the Operation Process include the following:

- *Concept Documents* – Concept documents generated early in the life cycle are used to direct the activities of this process
- *Operator/Maintainer Training*
- *Initial Trained Operators and Maintainers* – Operation of the system includes the humans that will operate, maintain, and sustain the system

- *Validated System*
- *Validation Report* – Including documentation of the validation activity results, a record of any recommended corrective actions, Design Feedback/Corrective Actions taken, and evidence that the system element or system satisfies the requirements, or not.

This process is governed by the following controls and enablers:

- *Applicable Laws and Regulations*
- *Industry Standards* – relevant industry specifications and standards
- *Agreements* – terms and conditions of the agreements
- *Project Procedures and Standards* – including project plans
- *Project Directives*
- *Organization/Enterprise Policies, Procedures, and Standards* – including guidelines and reporting mechanisms
- *Organization/Enterprise Infrastructure*
- *Project Infrastructure*
- *Operation Enabling Systems.*

4.9.1.4 Outputs

Outputs of the Operation Process include the following:

- *Operation Strategy* – Including staffing and sustainment of enabling systems and materials
- *Operation Enabling System Requirements* – Requirements for any systems needed to enable operation of the system-of-interest need to be developed
- *Operation Constraints on Design* – Any constraints on the design arising from the operation strategy to influence future design and specification of similar systems or reused systems-elements
- *Operation Report* – Including:
  - System performance reports (e.g., statistics, usage data, and operational cost data)
  - System trouble/anomaly reports – with recommendations for appropriate action.

4.9.1.5 Process Activities

The Operation Process includes the following activities:

- *Prepare for Operation*
- *Perform Operational Activation and Check-out*
  - Provide operator training and maintain qualified staff
- *Use System for Operations*
  - Execute ConOps for the system-of-interest
  - Track system performance and account for operational availability
  - Perform operational analysis
- *Perform Operational Problem Resolution*
  - Manage operational support logistics
  - Document system status and actions taken
  - Report malfunctions and make recommendations for improvement
- *Support the Customer.*

Common approaches and tips:

- Depending on the nature of agreements between different organizations, the development team may continuously or routinely communicate with users to determine the degree to which delivered services continue to satisfy their needs. The system may exhibit unacceptable performance when system elements implemented in hardware have exceeded their useful life or changes in the operational environment affect system performance. In the event of system failures or anomalies, it may be necessary to conduct engineering investigations to identify the source(s) of the failure and determine appropriate corrective actions. Systems engineers can assist in these activities.

4.9.2 Elaboration

Elaborations for this section are under development and will be included in a future revision of the *INCOSE Systems Engineering Handbook*.

4.10 Maintenance Process

4.10.1 Overview

4.10.1.1 Purpose

As stated in ISO/IEC 15288:2008:

> *The purpose of the Maintenance Process is to sustain the capability of the system to provide a service.*
>
> *This process monitors the system's capability to deliver services, records problems for analysis, takes corrective, adaptive, perfective and preventive actions and confirms restored capability.*[35]

4.10.1.2 Description

The Maintenance Process, as illustrated in Figure 4-19, includes the activities to provide operations support, logistics, and material management. Based on feedback from ongoing monitoring of the operational environment, problems are identified and corrective, remedial, or preventive actions are taken to restore full system capability. This process also contributes to the Requirements Analysis Process when considerations of constraints imposed in later life-cycle stages are used to influence the system requirements and architectural design.

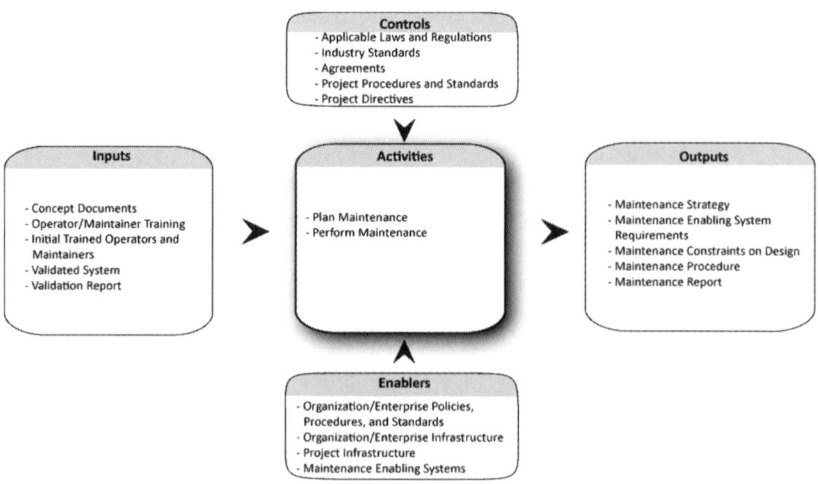

Figure 4-19 Context Diagram for the Maintenance Process

4.10.1.3 Inputs

Inputs to the Maintenance Process include the following:

- *Concept Documents* – Concept documents generated early in the life cycle are used to direct the activities of this process
- *Operator/Maintainer Training*
- *Initial Trained Operators and Maintainers* – Operation of the system includes the humans that will operate, maintain, and sustain the system
- *Validated System*
- *Validation Report* – Including documentation of the validation activity results, a record of any recommended corrective actions, Design Feedback/Corrective Actions taken, and evidence that the system element or system satisfies the requirements, or not.

This process is governed by the following controls and enablers:

- *Applicable Laws and Regulations*
- *Industry Standards* – relevant industry specifications and standards
- *Agreements* – terms and conditions of the agreements
- *Project Procedures and Standards* – including project plans
- *Project Directives*
- *Organization/Enterprise Policies, Procedures, and Standards* – including guidelines and reporting mechanisms
- *Organization/Enterprise Infrastructure*
- *Project Infrastructure*
- *Maintenance Enabling Systems.*

4.10.1.4 Outputs

Outputs of the Maintenance Process include the following:

- *Maintenance Strategy* – Accounts for the system's technical availability, replacements for system elements and logistical support, maintenance personnel training and staff requirements
- *Maintenance Enabling System Requirements* – Requirements for any systems needed to enable maintenance of the system-of-interest need to be developed

- *Maintenance Constraints on Design* – Any constraints on the design arising from the maintenance strategy
- *Maintenance Procedure*
- *Maintenance Report* – Including documentation of the maintenance activity results, reporting of failures and recommendations for action, and failure and lifetime performance data. This report also documents any required procedure or system changes that should be accomplished as part of on-going configuration management activities.

4.10.1.5 Process Activities

The Maintenance Process includes the following activities:

- *Plan Maintenance*
  - Establish a maintenance strategy
  - Define maintenance constraints on the system requirements
  - Obtain the enabling systems, system elements, and other services used for maintenance of the system
  - Monitor replenishment levels of spare parts
  - Manage the skills and availability of trained maintenance personnel
- *Perform Maintenance*
  - Implement maintenance and problem resolution procedures – including scheduled replacement of system elements prior to failure (i.e., preventive maintenance)
  - Maintain a history of failures, actions taken, and other trends to inform operations and maintenance personnel and other projects creating or utilizing similar system elements
  - Monitor customer satisfaction with system and maintenance support.

Common approaches and tips:

- Use historic data and performance statistics to maintain high levels of reliability and availability and to provide input to improve the design of operational and future systems.
- Planning for maintenance begins early in the system life cycle with the development of supportability criteria. These criteria, which include

reliability and maintainability requirements as well as personnel, training, facilities, etc., are included in the defined stakeholder requirements or system specification to ensure that they are considered in the system design.

- Maintain configuration management control throughout the Utilization and Support Stages in support of the Maintenance Process.

4.10.2 Elaboration

Elaborations for this section are under development and will be included in a future revision of the *INCOSE Systems Engineering Handbook*.

4.11 Disposal Process

4.11.1 Overview

4.11.1.1 Purpose

As stated in ISO/IEC 15288:2008:

> *The purpose of the Disposal Process is to end the existence of a system entity.*
>
> *This process deactivates, disassembles and removes the system and any waste products, consigning them to a final condition and returning the environment to its original or an acceptable condition. This process destroys, stores or reclaims system entities and waste products in an environmentally sound manner, in accordance with legislation, agreements, organizational constraints and stakeholder requirements. Where required, it maintains records in order that the health of operators and users, and the safety of the environment, can be monitored.*[36]

The Disposal Process is conducted in accordance with applicable guidance, policy, regulations, and statutes.

4.11.1.2 Description

Disposal is a life-cycle support process because concurrent consideration of disposal during the Development Stage generates requirements and constraints that must be balanced with defined stakeholders' requirements and other design considerations. Further, environmental concerns are driving the designer to consider reclaiming the materials or recycling them into new systems. Figure 4-20 is the context diagram for the Disposal Process.

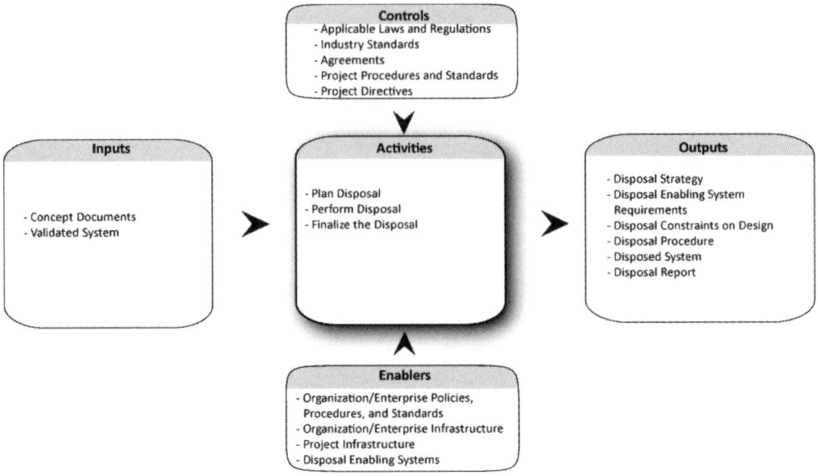

Figure 4-20 Context Diagram for the Disposal Process

4.11.1.3 Inputs

Inputs to the Disposal Process include the following:

- *Concept Documents* – Concept documents generated early in the life cycle are used to direct the activities of this process.
- *Validated System* – The Disposal Process works on a depleted system of system elements (e.g., batteries) meaning that if production and operational environments must be restored to former conditions, details of the initial state are relevant.

This process is governed by the following controls and enablers:

- *Applicable Laws and Regulations*
- *Industry Standards* – relevant industry specifications and standards
- *Agreements* – terms and conditions of the agreements
- *Project Procedures and Standards* – including project plans
- *Project Directives*
- *Organization/Enterprise Policies, Procedures, and Standards* – including guidelines and reporting mechanisms
- *Organization/Enterprise Infrastructure*
- *Project Infrastructure*
- *Disposal Enabling Systems.*

4.11.1.4 Outputs

Outputs of the Disposal Process include the following:

- *Disposal Strategy*
- *Disposal Enabling System Requirements* – Requirements for any systems needed to enable disposal of the system-of-interest need to be developed
- *Disposal Constraints on Design* – Any constraints on the design arising from the disposal strategy
- *Disposal Procedure*
- *Disposed System*
- *Disposal Report* Including documentation of the disposal activity results, may include an inventory of system elements for reuse/storage and any documentation or reporting required by regulation or organization standards.

4.11.1.5 Process Activities

The Disposal Process includes any steps necessary to return the environment to an acceptable condition; handle all system elements and waste products in an environmentally sound manner in accordance with applicable legislation, organizational constraints, and stakeholder agreements; and document and retain records of Disposal activities, as required for monitoring by external oversight or regulatory agencies. In general, Disposal Process includes the following activities:

- *Plan Disposal*
  - Review the Concept of Disposal, including any hazardous materials and other environmental impacts to be encountered during disposal
  - Define the Disposal Strategy
  - Impose associated constraints on the system requirements
- *Perform Disposal*
  - Deactivate the elements to be terminated
  - Disassemble the elements for ease of handling
  - Remove the elements and any associated waste products from the operational site – includes removing materials from storage

sites and consigning the elements and waste products for destruction or permanent storage

- *Finalize the Disposal*
  - Maintain documentation of all Disposal activities and residual hazards.

Common approaches and tips:

- The project team conducts analyses to develop solutions for ultimate disposition of the system, constituent elements, and waste products based on evaluation of alternative disposal methods available. Methods addressed should include storing, dismantling, reusing, recycling, reprocessing, and destroying end products, enabling systems, system elements, and materials.
- Disposal analyses include consideration of costs, disposal sites, environmental impacts, health and safety issues, responsible agencies, handling and shipping, supporting items, and applicable federal, state, local, and host-nation regulations.
- Disposal analyses support selection of system elements and materials that will be used in the system design, and should be readdressed to consider design and project impacts from changing laws and regulations throughout the project life cycle.
- Disposal Strategy and design considerations are updated throughout the system life cycle in response to changes in applicable laws, regulations, and policy.
- Consider donating an obsolete system. – Many items, both systems and information, of cultural and historical value have been lost to posterity because museums and conservatories were not considered as an option during the disposal stage.
- Concepts such as Zero Footprint and Zero Emissions drive current trends toward corporate social responsibility that influence decision-making regarding cleaner production and operational environments and eventual disposal of depleted materials and systems.[37]
- The ISO 14000 series includes standards for Environmental Management Systems and Life-Cycle Assessment.[38]
- Instead of designing cradle-to-grave products, dumped in landfills at the end of their 'life,' a new concept is transforming industry by creating products for cradle-to-cradle cycles, whose materials are

perpetually circulated in closed loops. Maintaining materials in closed loops maximizes material value without damaging ecosystems.[39]

4.11.2 Elaboration

Elaborations for this section are under development and will be included in a future revision of the *INCOSE Systems Engineering Handbook*.

4.12 Cross-Cutting Technical Methods

The previous sections provided a serial description of the Technical Processes used across the system life cycle. This section provides insight into methods that cut across the Technical Processes, reflecting various aspects of the iterative and recursive nature of SE.

4.12.1 Modeling, Simulation, and Prototyping

Designers have used models and simulations for centuries both to check their own thinking and to communicate their concepts to others. The benefit is two-fold: (1) models and simulations confirm anticipated system behaviors before proceeding with development of an actual system, and (2) models and simulations presenting a clear, coherent design to those who will develop, test, deploy, and evolve the system, thereby maximizing productivity and minimizing error. The ability to detect faults and incompatibilities via system models and simulations early in a project helps avoid higher project cost and schedule overruns later in a project, especially during system operation.

The objective of modeling and simulation is to obtain information about the system before significant resources are committed to its design, development, construction, verification, or operation. To that end, modeling and simulation helps generate data in the domain of the analyst or reviewer, not available from existing sources, in a manner that is affordable and timely to support decision-making. An adequate, accurate, and timely models and simulations inform stakeholders of the implications of their preferences, provide perspective for evaluating alternatives, and build confidence in the effects that an implemented system will produce. They also help the development, deployment, and operational staffs comprehend the design requirements, appreciate imposed limits from technology and management, and ensure an adequate degree of sustainability. Finally, an adequate, accurate, and timely models and simulations help the organization and its suppliers and to provide the necessary and sufficient personnel, methods, tools, and infrastructure for system realization.

The long-term benefits of modeling and simulation are commensurate with the gap between the extent, variety, and ambiguity of the problem and the competencies of downstream staffing. A relatively simple model of an intended system may be sufficient for a highly competent staff whereas a much more elaborate simulation may be necessary for a less competent staff, especially one faced with producing a novel, large-scale system that is capable of autonomously coping with unpredictable mission situations. Ultimately, the benefit of modeling and simulation is proportional to the stakeholders' perception of the timeliness, trustworthiness, and ease of use and maintenance of the model or simulation. Consequently, the planned resources anticipated to be spent in development, validation, and operation of the model must be consistent with the expected value of the information obtained through use of the model.

General Principles

Models represent the essential characteristics of the system under development, the environment in which the system operates, and/or interactions with enabling systems and interfacing systems. Models can be used within most systems life-cycle processes, for example:

- *Intended usage thus affects MOEs* – Descriptive model of the problematic situation ensures the right problem(s) are being addressed
- *Requirements Analysis* – Enables justification of requirements and avoids over/underspecification
- *Architectural Design* – Evaluate candidate options against selection criteria and enable active agents to discover the best architecture
- *Design & Development* – Obtain needed design data, adjust parameters for optimization, and update system model fidelity as actual data for components become available
- *Verification* – Simulate the system's environment, evaluate verification data (simulation uses observable data as inputs for computation of critical parameters that are not directly observable), and validate the fidelity of the simulation (false positives/false negatives)
- *Operations* – Update fidelity of simulation to reflect actual behavior and simulate operations in advance of execution for planning and validation.

Most systems start as a primitive model that is elaborated and translated in several stages. A model may be manifest as sketches, textual specifications, graphics/images, mockups, scale models, prototypes, and/or emulations. Often, separate models are prepared for distinct viewpoints, such as functional, performance, reliability, survivability, operational availability, and cost. The result of modeling is a prediction of characteristics (i.e., performance, reliability, operations, and cost, etc.) across the spectrum of system attributes throughout its life cycle. The predictions are used to guide decisions about the system's design, construction, and operation, or to verify its acceptability. Standard tools for all types of modeling are available commercially for a wide range of system characteristics.

Many models employ some logical or mathematical rule to convert a set of inputs to corresponding outputs with the same form of dependence as in the represented system, but do not mimic the structure of the system. Two types of models are particularly important to SE: descriptive and prescriptive. Descriptive models clarify the context, content, structure, and behavior of the problem; identify what stakeholders are willing to do about the problem; and describe the relevant characteristics of appropriate technologies. Prescriptive models specify the intended system in terms of its necessary and sufficient capabilities, the order of all relevant technologies and components, and the emergent characteristics expected of the system throughout operational situations, modes, and phases. Prescriptive models also contain or generate an estimate of the likelihood that an envisioned system will meet its MOEs. Validity depends on showing, through analysis or empirical data, that the representation tracks the actual system in the region of concern.

Simulations can reflect system functions or the detailed structure of the system. They are composed of representations of system elements, connected in the same manner as in the actual system. Usually the simulation is run through scenarios in the time domain to simulate the behavior of the real system. An example might be the simulation of a fluid control system made up of representations of the piping, pump, control valve, sensors, control circuit, and the fluid running through the system.

A simulation can also be used to quickly examine a range of sizes and parameters, not just a "Point Design." This will help ensure that the "best" solution is obtained. For example, does the system have the proper size throughout, with no choke points? Simulations use scenarios extracted from the ConOps with inputs based on system requirements. Monte Carlo runs may be made to get averages and probability distributions. In addition to examining nominal conditions, off-nominal runs should also be made to establish system reactions or breakage when exposed to extraordinary (out-of-spec) conditions.

Types of Models and Simulations

Systems Engineering models and simulations typically reflect four paradigms:

- Functional analysis with specialty, environmental, and interface engineering models attached (see Section 4.12.2)

- Modern Structured Analysis/Process for System Architecture and Requirements Engineering
- Systems Modeling Language (SysML™; see Section 4.12.3)
- Context-sensitive systems, also called complex, adaptive systems, which modify their internal gradients, architecture, and content depending on interactions with their environment.

Each of these sets can be applied comprehensively to a problem no matter how the system is to be implemented (e.g., hardware, software, people). This comprehensive application has been referred to as the *Universal Architecture Description Framework*.[40]

The specific type of model or simulation selected depends on the particular characteristics of the system that are of interest. Generally, it focuses on some subset of the total system characteristics, such as timing, process behavior, or various performance measures.

Models may be made up of one or several of the following types:

- Physical (e.g., Wind Tunnel model, Mockups, Acoustic model, structural test model, engineering model, prototypes)
- Graphical (e.g., N^2 charts, Behavior diagrams, Program Evaluation Review Technique [PERT] charts, Logic Trees, blueprints)
- Mathematical (deterministic; e.g., Eigen value calculations, Dynamic motion, Cost)
- Statistical (e.g., Monte Carlo, Process modeling, sequence estimation).

Physical models exist as tangible, real-world objects that are identical or similar in the relevant attributes to the actual system. The physical properties of the model are used to represent the corresponding properties of the actual system. Examples of physical models include: wind tunnels, test beds, and breadboards/brassboards.

Graphical models are a mapping of the relevant attributes of the actual system onto a graphical entity with analogous attributes. The geometric or topological properties of the graphical entity are used to represent geometric properties, logical relationships, or process features of the actual system. Examples of graphical models include: FFBDs, N^2 diagrams, logic trees, blueprints, schematics, and maps.

Mathematical (deterministic) models use closed mathematical expressions or numerical methods to convert input data to outputs with the same functional dependence as the actual system. Mathematical equations in closed or open form are constructed to represent the system. The equations are solved using appropriate analytical or numerical methods to obtain a set of formulae or tabular data defining the predicted behavior of the system. Examples of mathematical models include: operational or production throughput analysis, thermal analysis, vibration analysis, load analysis, stress analysis, Eigen value calculations, and linear programming.

Statistical models are used to generate a probability distribution for expected outcomes, given the input parameters and data. Statistical models are appropriate whenever truly random phenomena are involved (as with reliability estimates) whenever there is uncertainty regarding the inputs such that the input is represented by a probability distribution, or whenever the collective effect of a large number of events may be approximated by a statistical distribution. Examples of statistical models include: Monte Carlo, logistical support, discrete, and continuous models.

Most current models describe static systems and allude to a limited spectrum of dynamics and behaviors. Far more beneficial models enable exploration of how well and to what extent the intended system responds to external and internal change by adjusting its gradients, altering its pattern of internal relationships, or aligning its content. Executable models enable emulation of operational modes (e.g., nominal, degraded, recovery, training, maintenance, diagnostic) and help establish the viability of SoS scenarios and engineering change proposals (ECPs). Wymorean models[41], for example, clarify six aspects of a system: input/output, performance, technology, cost, test, and tradeoff gradients.

Operational Evaluation Modeling for Context-Sensitive Systems (OpEMCSS) models is context sensitive, which means that systems engineers can model two or more implicit systems as a whole system and execute the model by selecting scenarios that represent the system context. OpEMCSS models can contain intelligent agents that can continuously adjust gradients, alter component interoperation, and align content to maximize stated MOEs of the sponsored system. Axiom-based models also apply intelligent agents.

Steps in Modeling and Simulation

The general steps in the application of modeling and simulation are as follows:

1. *Select the appropriate type(s) of model (or simulation)* (e.g., Quick, Economical, or Accurate).

2. *Design the model (or simulation)* – Care is needed in the design of the model to ensure that the general criteria are met. Usually this requires some degree of fundamental analysis of the system:

    - Identify the relevant system characteristics that are to be evaluated through use of the model.
    - Determine the relevant measurable parameters that define those characteristics, and separate them from irrelevant parameters.
    - Define the scope and content of data needed to support the decision economically and accurately.

    It is particularly important that the model be economical in the use of time and resources, and that the output data be compact and readily understandable to support efficient decisions. The Taguchi Design of Experiments process—which identifies the sensitivity of the results to variation of key parameters and adjusts the spacing of sampling so that the total range of results is spanned with the minimum number of verification points—can be very effective in determining the bounds and the limits of the model. This data can be used to estimate the value of the information gained by producing the model.

    The model itself can be considered as a system to which the Requirements Analysis and Architectural Design (i.e., functional analysis and system synthesis) steps of the SE process are applied to determine the requirements for the model and define the approach. This analysis provides an overall description of the modeling approach. Following its review and approval, the detailed definition of the model can be created according to usual practice for the type of model selected.

3. *Validate the model (or simulation)* through an appropriate method to the satisfaction of responsible parties – It is crucial to prove that the model is trustworthy and suitably represents reality, particularly in cases where a feel for system behavior is absent, or when serious consequences can result from inaccuracy. Models can be validated by:

    - Determining "fit for purpose," as described by Friedman[42]
    - Experience with application of similar models in similar circumstances
    - Analysis showing that the elements of the model are of necessity correct and are correctly integrated

- Comparison with test cases in the form of independent models of proven validity or actual test data
- The modeling schema itself can be validated by using small-scale models.

4. *Document the model (or simulation),* including background, development process, a complete description of the model itself and its validation, and a record of activities and data generated by its use, sufficient to support evaluation of model results and further use of the model.
5. *Obtain needed input data and operate the model (or simulation)* to obtain desired output data to represent the actual system and its operating environment – In some situations, defining and acquiring the basis model data can be a very large effort, so care in design of the model is needed to minimize this problem. Perform as many runs as are needed to span the range of the system parameters and operating conditions to be studied, and in the case of statistical models, to develop the needed level of statistical validity.
6. *Evaluate the data* to create a recommendation for the decision in question.
7. *Review the entire process*, iterating as necessary to make corrections and improvements, to ensure that the model supports the conclusions reached – Explore the sensitivity of the result to changes in initial assumptions, data, and processes. If the result has an adequate level of confidence in an unambiguous decision, then the task is complete. Otherwise, look for corrections or improvements to the process and iterate.
8. *Evolve the model (or simulation), as necessary* – In some cases, a model, created initially to support analysis of the system, evolves to become a deliverable portion of the system. This can occur in cases such as a model of system dynamics, which then becomes the core of the system control system, or an operations simulation model, which evolves into a tool for system operations planning used in the operational stage. The potential for the model to evolve in this manner should be a factor in initial selection and design of the model; anticipation of future uses of the model should be included in its initial conception.

4.12.1.1 Prototyping

Prototyping is a technique that can significantly enhance the likelihood of providing a system that will meet the user's need. In addition, a prototype can facilitate both the awareness and understanding of user needs and stakeholder requirements. Two types of prototyping are commonly used: rapid and traditional.

Rapid prototyping is probably the easiest and one of the fastest ways to get user performance data and evaluate alternate concepts. A rapid prototype is a particular type of simulation quickly assembled from a menu of existing physical, graphical, or mathematical elements. Examples include tools such as laser lithography or computer simulation shells. They are frequently used to investigate form and fit, human-system interface, operations, or producibility considerations. Rapid prototypes are widely used and are very useful; but, except in rare cases, they are not truly "prototypes."

Traditional prototyping is a tool that can reduce risk or uncertainty. A partial prototype is used to verify critical elements of the system-of-interest. A full prototype is a complete representation of the system. It must be complete and accurate in the aspects of concern. Objective and quantitative data on performance times and error rates can be obtained from these higher fidelity interactive prototypes.

The original use of a prototype was as the first-of-a-kind product from which all others were replicated. However, prototypes are not "the first draft" of production entities. Prototypes are intended to enhance learning and should be set aside when this purpose is achieved. Once the prototype is functioning, changes will often be made to improve performance or reduce production costs. Thus, the production entity may require different behavior. The Maglev train system (see Section 3.6.3) may be considered a prototype (in this case, proof-of-concept) for longer distance systems that will exhibit some but not all of the characteristics of the short line. Scientists and engineers are in a much better position to evaluate modifications that will be needed to create the next system because of the existence of a traditional prototype.

4.12.2 Functions-Based Systems Engineering Method

4.12.2.1 Overview

A *function* is a characteristic task, action, or activity that must be performed to achieve a desired outcome. A function may be accomplished by one or more system elements comprised of equipment (hardware), software, firmware, facilities, personnel, and procedural data.

Functional Analysis/Allocation is an examination of a defined function to identify all the sub-functions necessary to accomplish that function; all usage modes must be included in the analysis. This activity is conducted to the level of depth needed to support required synthesis efforts. Identified functional requirements are analyzed to determine the lower-level functions required to accomplish the parent requirement. Every function that must be performed by the system to meet the operational requirements is identified and defined in terms of allocated functional, performance, and other limiting requirements. Each function is then decomposed into sub-functions, and the requirements allocated to the function are each decomposed with it. This process is iterated until the system has been completely decomposed into basic sub-functions and each sub-function at the lowest level is completely, simply, and uniquely defined by its requirements. In the process, the interfaces between each of the functions and sub-functions are fully defined, as are the interfaces to the external world.

Identified sub-functions are arranged in a functional architecture to show their relationships and interfaces (internal and external). Functional requirements should be arranged in their logical sequence so that lower-level functional requirements are recognized as part of higher-level requirements. Functions should have their input, output, and functional interface requirements (both internal and external) defined and be traceable from beginning to end conditions. Time critical requirements must also be analyzed.

Performance requirements should be successively established, from the highest to lowest level, for each functional requirement and interface. Upper-level performance requirements are then flowed down and allocated to lower-level sub-functions. Timing requirements that are prerequisite for a function or set of functions must be determined and allocated. The resulting set of requirements should be defined in measurable terms and in sufficient detail for use as design criteria. Performance requirements should be traceable from the lowest level of the current functional architecture, through the analysis by which they were allocated, to the higher-level requirement they are intended to support. All of these types of product requirements must also be verified.

Functional analysis/allocation should be conducted iteratively:

- To define successively lower-level functions required to satisfy higher-level functional requirements and to define alternative sets of functional requirements.
- With requirements analysis, to define mission and environment driven performance and to determine that higher-level requirements are satisfied.
- To flow down performance requirements and design constraints.
- With design synthesis, to refine the definition of product and process solutions.

Purpose

The objective of Functional Analysis/Allocation is to create a functional architecture for which system products and processes can be designed and to provide the foundation for defining the system architecture through the allocation of functions and sub-functions to hardware/software, databases, facilities and operations (i.e., personnel). It does not describe either the hardware architecture or software architecture of the system; those architectures are developed during System Synthesis in the Architectural Design portion of the SE process.

Functional Analysis/Allocation describes what the system will do, not how it will do it. Ideally, this process begins only after all of the system requirements have been fully identified. This means that Requirements Analysis (see Section 4.1.2.6) must be completed before this process starts. Often, this will not be possible, and these tasks will have to be done iteratively, with the functional architecture being further defined as the system requirements evolve.

Inputs

Representative inputs to Functional Analysis/Allocation are as follows:

- Functional requirements
- Performance requirements
- Program decision requirements (such as objectives to reuse certain hardware and software or use COTS items)
- Specifications and Standards requirements
- Architectural concepts
- ConOps
- Constraints.

Outputs

The products of the Functional Analysis/Allocation process can take various formats depending on the specific stage of the process and on the specific technique used to develop the functional architecture. The following are some key outputs generated from the functional analysis activity:

1. *Behavior Diagrams* – Describe behavior that specifies system-level stimulus responses using constructs that specify time sequences, concurrencies, conditions, synchronization points, state information and performance.

2. *Context Diagrams* – Top-level diagram of a data flow that is related to a specific level of system decomposition. This diagram portrays all inputs and outputs of a system but shows no decomposition.

3. *Control Flow Diagrams* – Depict the set of all possible sequences in which operations may be performed by a system or a software program. There are several types of Control Flow Diagrams, including box diagrams, flowcharts, input-process-output charts, and state transition diagrams.

4. *Data Flow Diagrams* – Provide an interconnection of each of the behaviors that the system must perform. All inputs to the behavior designator and all outputs that must be generated are identified along with each of the data stores that each must access. Each of the Data Flow Diagrams must be checked to verify consistency with the Context Diagram or higher level Data Flow Diagram.

5. *Data Dictionaries* – Documentation that provides a standard set of definitions of data flows, data elements, files, etc. as an aid to communications across the development organizations.

6. *Entity Relationship Diagrams* – Depict a set of entities (e.g., functions or architecture elements) and the logical relationships between them.

7. *Functional Flow Block Diagrams (FFBD)* – Relate the inputs and outputs and provide some insight into flow between the system functions.

8. *Models* – Abstractions of relevant characteristics of a system used as a means to understand, communicate, design, and evaluate (including simulation) a system. They are used before the system is built and while it is being verified or in service. Section 4.3.2.6, *Model, Simulate, and Prototype System Architectures*, states:

   > *Modeling, simulation, and prototyping used during architecture design can significantly reduce the risk of failure in the finished system [...]. These techniques enable the development of complex and costly enabling systems. Systems engineers use modeling and simulation on large complex projects to manage the risk of failure to meet system mission and performance requirements.*

9. *Simulation Results* – Output from a model of the system that behaves or operates like the system under interest when provided a set of controlled inputs.

10. *Integrated Definition for Functional Modeling (IDEF) Diagrams* – Process control diagrams that show the relationship between functions by sequential input and output flows. Process controls enter the top of each represented function, and lines entering the bottom show the supporting mechanism needed by the function.

These various outputs characterize the functional architecture. There is no one preferred output tool that will support this analysis. In many cases, several of these are necessary to understand the functional architecture and the risks that may be inherent in the subsequent synthesis of system architecture. Using more than one of these formats allows for a "check and balance" of the analysis process and will aid in communication across the system design team.

Process Activities

The Functional Analysis/Allocation process is iterative, even within a single stage in the system life-cycle. The functional architecture begins at the top level as a set of functions that are defined in the applicable requirements document or specification, each with functional, performance, and limiting requirements allocated to it (in the extreme, top-level case, the only function is the system, and all requirements are allocated to it). As shown in Figure 4-21, the next lower level of the functional architecture is developed and evaluated to determine whether further decomposition is required. If it is, then the process is iterated through a series of levels until a functional architecture is complete. If not, then the process is complete and System Synthesis can begin.

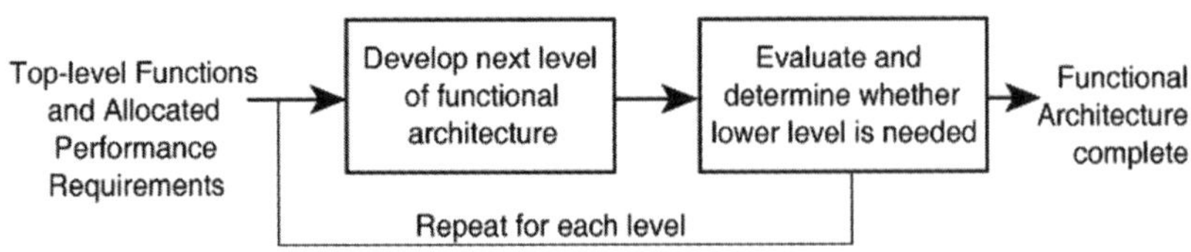

Figure 4-21 Functional Analysis/Allocation Process

At each level of the process, alternative decompositions and allocations may be considered and evaluated for each function and a single version selected. After all of the functions have been identified, then all the internal and external interfaces to the decomposed sub-functions are established. These steps are shown in Figure 4-22.

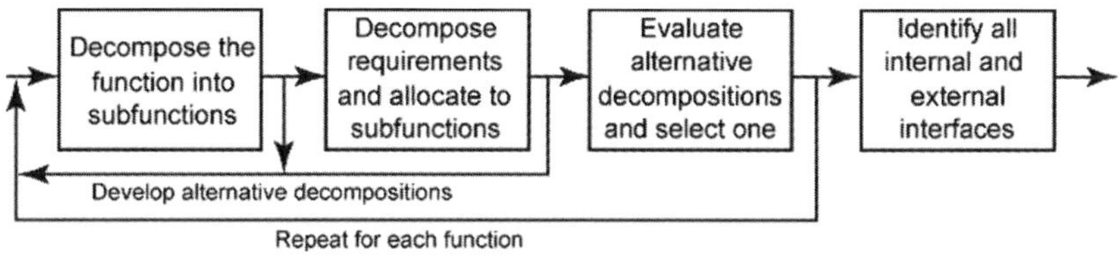

Figure 4-22 Alternative Functional Decomposition Evaluation and Definition

Note that while performance requirements may be decomposed and allocated at each level of the functional decomposition, it is sometimes necessary to proceed through multiple levels before allocating the performance requirements. Also, sometimes it is necessary to develop alternative functional architectures and conduct a trade study to determine a preferred one. With each iteration of Functional Analysis/Allocation, alternative decompositions are evaluated and all interfaces are defined.

Tools that can be used to perform *Functional Analysis/Allocation* include:

- Analysis tools
- Modeling and Simulation tools
- Prototyping tools
- Requirements traceability tools.

The INCOSE web site contains a current listing of applicable tools.

The following measures can be used to measure the overall process and products of *Functional Analysis/Allocation*:

1. Number of allocation-related trade studies completed as a percent of the number identified
2. Percent of analyses completed
3. Number of functions without a requirements allocation
4. Number of functions not decomposed
5. Number of alternative decompositions
6. Number of internal and external interfaces not completely defined
7. Depth of the functional hierarchy as a percentage versus the target depth
8. Percent of performance requirements that have been allocated at the lowest level of the functional hierarchy.

4.12.2.2 Elaboration

Decompose Each Function to Lower-Level Functions

A function is accomplished by one of many system or subsystem elements, namely, equipment, software, facilities, or personnel. The stepwise decomposition of a system can be viewed as a top-down approach to problem solving. Functional identification and decomposition can be performed with respect to logical groupings, time ordering, data flow, control flow, state transitions, or some other criterion.

The objective of the functional decomposition activity is to develop a hierarchy of FFBDs that meet all the functional requirements of the system. Note, however, that this hierarchy is only a portion of the functional architecture. The architecture is not complete until all of the performance and limiting requirements have been appropriately decomposed and allocated to the elements of the hierarchy, as described below.

For the initial iteration of *Functional Analysis/Allocation*, the baseline requirements and operational concept have been identified during *Requirements Analysis*. First, top-level system functions are determined by evaluating the total set of baseline requirements as they map to the system-level design, keeping in mind the desire to have highly cohesive, loosely coupled functions. The result is a set of top-level functions that, when appropriately grouped together, provide the required capabilities of each element in the system-level design. Each of the top-level functions is then further refined to lower-level functions based on its associated requirements.

Decomposition of the function involves the creation of a network of lower-level "child" functions, each of which receives its allocated portion of the "parent's" functional requirements. In this process, each functional requirement is decomposed into lower-level requirements, and each of these is allocated to a lower-level function (i.e., sub-function) in the next-level FFBD. Functional interfaces fall out of this process.

A description of each function in the hierarchy should be developed to include the following:

1. Its place in a network (e.g., FFBD or IDEF0/1 diagrams) characterizing its interrelationship with the other functions at its level
2. The set of functional requirements that have been allocated to it and define what it does
3. Its inputs and outputs, both internal and external.

This activity may use various graphical methods to capture the results of the analysis, including structured analysis, such as Data Flow Diagrams, IDEF0/1 diagrams, Control Flow Diagrams, or other modern techniques. These are all forms of the Functional Descriptions.

In undertaking the Functional Analysis/Allocation process, it is important to establish criteria for completing (or stopping) the functional decomposition. Usually, this means continuing until the functional requirement is clear and realizable in hardware, software, and/or manual operations. In some cases, the engineer will continue the effort beyond what is necessary until funding for the activity has been exhausted. In establishing the "stop" criteria, recognize that the objective of pushing the decomposition to greater detail is to reduce program risk. At some point, the incremental risk reduction becomes smaller than the cost (in both time and money) of the effort to further decompose. Each program will be different, so it is impossible to set forth all-purpose stop criteria. The program manager and Systems Engineer who understand their specific program's risks need to establish their own stop criteria early in the process and ensure that the decomposition efforts are frequently reviewed.

Allocate Performance and Other Limiting Requirements

Requirements allocation is the further decomposition of system-level requirements until a level is reached at which a specific hardware item or software routine can fulfill the needed functional/performance requirements. It is the logical extension of the initial functional identification and an integral part of any functional analysis effort.

Functional requirements are fully allocated to functions and sub-functions in the decomposition step. The objective of this step is to have every performance or limiting requirement allocated to a function or sub-function at the next level in the hierarchy of FFBDs. Some performance requirements will have already been decomposed to do this, but additional requirements may need to be derived.

Some straightforward allocation of functional requirements can be made, but the procedure may involve the use of supporting analyses and simulations to allocate system-level requirements. An example of the need for additional analysis is the allocation of availability goals to CIs. These goals can only be expressed as maintainability and reliability requirements. Allocations are made by these parameters (e.g., maintainability and reliability), but only in conjunction with analytical and/or computer simulation to ascertain the impact of a given set of allocations on system availability.

If a requirement cannot be allocated as a single, independent entity, then it must be decomposed and the derived requirements allocated. Often this step requires some anticipation of the results of System Architecture Synthesis because decomposition of response-time or noise-level requirements is equivalent to developing timing or noise budgets. In some cases, it will be necessary to defer decomposition of performance and limiting requirements until multiple stages of the functional hierarchy have been developed.

Design constraints recognize inherent limitations on the sizing and capabilities of the system, its interfacing systems, and its operational and physical environment. These typically include power, weight, propellant, data throughput rates, memory, and other resources within the vehicle or which it processes. These resources must be properly managed to ensure mission success and that all constraints are identified to the designer prior to start of detailed design. This should prevent the need for redesign due to unidentified constraints. All SE groups should be involved in identifying and managing constraints, primarily the Engineering Specialties: Reliability, Maintainability, Producibility, Human Engineering, ElectroMagnetic Interference (EMI)/ ElectroMagnetic Compatibility (EMC), System Safety, Survivability, Support, Security, and LCC/Design-to-Cost.

Design constraints are of paramount importance in the development of derivative systems. A derivative system is a system that by mandate must retain major elements of a prior system. For example, an aircraft may be modified to increase its range while retaining its fuselage or some other major elements. Constraints must therefore be firmly established, such as: Which elements *must* remain unmodified?; What can be added?; and What can be modified? The key principle to be invoked in the development of derivative systems is that the requirements for the system as a whole must be achieved while conforming to the imposed constraints.

Care must be exercised such that the myriad engineering specialty requirements and constraints are incorporated into appropriate specifications. Incorporation of engineering specialties personnel into the Systems Engineering and Integration Team (SEIT) of an IPPD organization or into all appropriate IPDTs are ways of ensuring that their requirements are incorporated into specifications.

The following tasks constitute the bulk of the performance and other limiting requirements allocation activity:

1. Identify from the SOW all design constraints placed on the program. This particularly includes those from compliance documents.
2. Identify the groups defining constraints and incorporate them into the SE effort.

3. Analyze the appropriate standards and lessons learned to derive requirements to be placed on the hardware and software CI design.
4. Tailor the compliance documents to fit overall program needs.
5. Identify the cost goals allocated to the design.
6. Define system interfaces and identify or resolve any constraints that they impose.
7. Identify any COTS or NDI CIs that must be used and the constraints that they may impose.
8. Document all derived requirements in specifications and ensure that they are flowed down to the lowest CI level.
9. Ensure that all related documents (i.e., operating procedures, etc.) observe the appropriate constraints.
10. Review the design as it evolves to ensure compliance with documented constraints.

Evaluate Alternative Functional Decompositions and Select One

Not all functional decompositions are of equal merit. As such, it is necessary to consider alternative decompositions at each level and select the most promising. Because of the reality of system design constraints, target costs, or COTS/NDI elements, it is often desirable to produce multiple alternative functional architectures that can then be compared in a trade study to pick the one most effective in meeting the objectives.

Eventually, each sub-function in the lowest levels of the functional architecture is going to be allocated to hardware, software, or manual operations. In addition, each of these functions will have to be verified. The objective here is to select those decompositions that lend themselves to straightforward implementation and verification. Systems engineers may also be able to come up with decompositions that allow a single function to be used at several places within the hierarchy, thereby simplifying development.

This activity requires sound engineering judgment since there are various ad hoc figures of merit that can be applied to evaluate alternative decompositions. The degree of interconnectivity among functions is one possible measure. Several measures for software-intensive systems can also be applied, such as high cohesion and low coupling. The systems engineer needs to be aware of opportunities for using NDI hardware and software. That means that a sub-function that has already been implemented in a compatible form on another system may be preferred to one that has not.

Define/Refine Functional Interfaces (Internal and External)

Each function requires inputs to operate, and the subsequent product of a function is an output. The objective of this step is to identify and document where within the FFBD each function (or sub-function) will obtain its required inputs and where it will send its outputs. Therefore, all internal and external interfaces and the nature of the flows through each interface must be completely identified and defined.

N^2 diagrams (see Figure 4-23) are a systematic approach to identify, define, tabulate, design, analyze, and document functional interfaces. These apply to systems interfaces, equipment (i.e., hardware) interfaces, or software interfaces. N^2 diagrams can also be used at later stages of the development process to analyze and document physical interfaces between system elements. An N^2 diagram is a visual matrix, which requires the user to generate complete definitions of all the system interfaces in a rigid bi-directional, fixed framework.

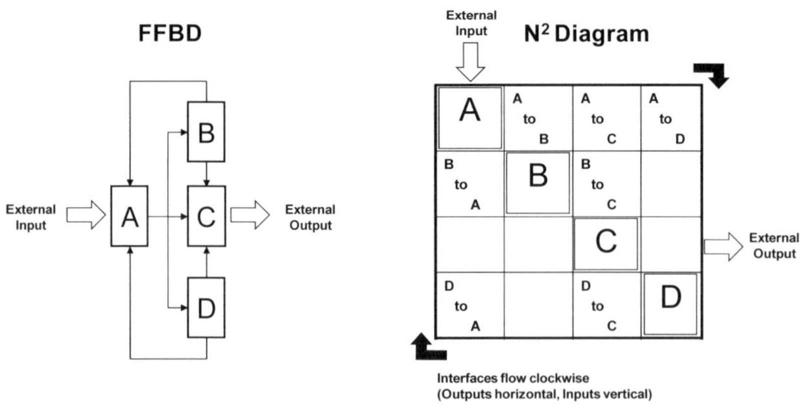

Figure 4-23 Sample FFBD and N^2 Diagram

The system functions are placed on the chart diagonal. The rest of the squares in the N by N matrix represent the interface inputs and outputs. Interfaces between functions flow in a clockwise direction. The entity being passed from Function A to Function B, for example, can be defined in the appropriate square. When a blank appears, there is no interface between the respective functions. When all functions have been compared to all other functions, then the chart is complete. If lower-level functions are identified in the process with corresponding lower-level interfaces, then they can be successively described in expanded or lower level diagrams. Sometimes characteristics of the entity passing between functions may be included in the box where the entity is identified. One of the main functions of the chart, besides interface identification, is to pinpoint areas where conflicts may arise between functions so that system integration later in the development cycle can proceed efficiently.[43,44,45]

Alternatively, or in addition, Data/Control Flow Diagrams can be used to characterize the flow of information among functions and between functions and the outside world. As the system architecture is decomposed to lower and lower levels, it is important to make sure that the interface definitions keep pace and that interfaces are not defined that ignore lower-level decompositions.

Define, Refine, and Integrate Functional Architecture

It may be necessary to make some final modifications to the functional definitions, FFBDs, and interfaces to arrive at a viable allocation. The product of this activity is a final FFBD hierarchy with each function (or sub-function) at the lowest possible level uniquely described. The functional flow diagrams, interface definitions, and allocation of requirements to functions and sub-functions constitute the system Functional Architecture.

4.12.3 Object-Oriented Systems Engineering Method[46]

4.12.3.1 Overview

The Object-Oriented Systems Engineering Method (OOSEM) integrates a top-down, model-based approach that can be used with the Object Management Group (OMG)'s System Modeling Language (SysML™) to support the specification, analysis, design, and verification of systems. OOSEM leverages object-oriented concepts in concert with more traditional top-down SE methods and other modeling techniques to help architect more flexible and extensible systems that can accommodate evolving technology and changing requirements. OOSEM is also intended to ease integration with object-oriented software development, hardware development, and verification.

Object-oriented SE evolved from work in the mid 1990's at the Software Productivity Consortium (now the Systems and Software Consortium) in collaboration with Lockheed Martin Corporation. The methodology was applied in part to a large distributed information system development at Lockheed Martin that included hardware, software, database, and manual procedure elements. INCOSE Chesapeake Chapter established the OOSEM Working Group in November 2000 to help further evolve the methodology. OOSEM is summarized in various industry and INCOSE papers[47,48] and in the book entitled, *A Practical Guide to SysML: Systems Modeling Language*, by Friedenthal, Moore, and Steiner.[49] An introduction to the methodology is also available in the full-day *Object-Oriented Systems Engineering Method (OOSEM)*

Tutorial, by the Lockheed Martin Corporation and INCOSE OOSEM Working Group.[50]

The OOSEM objectives are as follows:

- Capture and analyze requirements and design information to specify complex systems
- Integration of model-based systems engineering (MBSE) methods with object-oriented software, hardware, and other engineering methods
- Support system-level reuse and design evolution.

As stated above, OOSEM is a hybrid approach that leverages object oriented techniques and an SE foundation. It also introduces some unique techniques, as indicated in Figure 4-24.

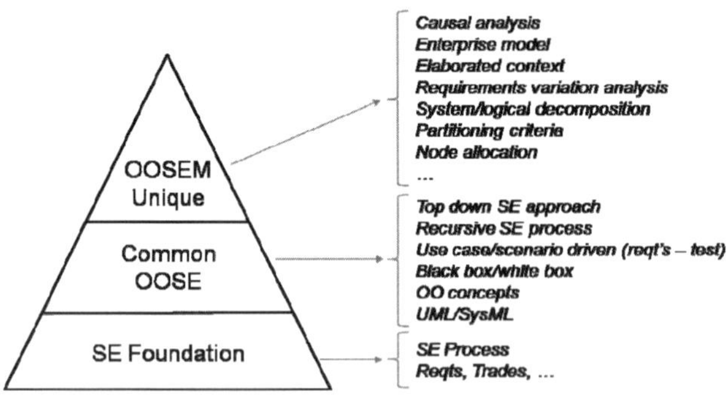

Figure 4-24 Foundation of OOSEM

The OOSEM supports a SE process as illustrated in Figure 4-25.

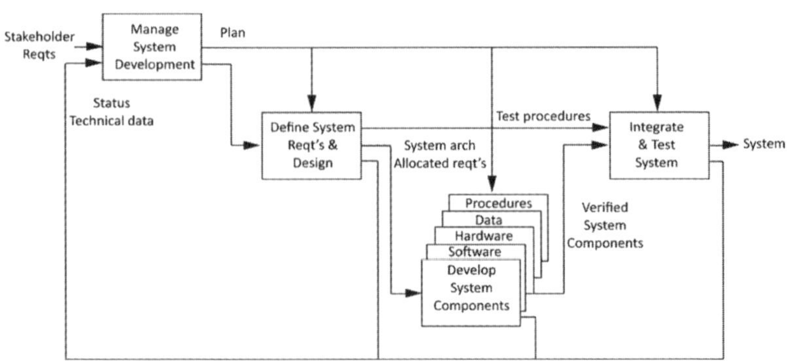

Figure 4-25 OOSEM Activities in the Context of the System Development Process

The core tenets of OOSEM include recognized practices essential to SE that include: (1) IPPD (see Section 5.1.2.3), essential to improve communications, and (2) a recursive "Vee" life-cycle process model that is applied to each level of the system hierarchy (see Section 3.3).

As shown in Figure 4-26, OOSEM includes the following development activities:

- Analyze Needs
- Define System Requirements
- Define Logical Architecture
- Synthesize Allocated Architectures
- Optimize and Evaluate Alternatives
- Validate and Verify System.

These activities are consistent with typical SE "Vee" processes that can be recursively and iteratively applied at each level of the system hierarchy. Fundamental tenets of SE, such as disciplined management processes (i.e., risk management, configuration management, planning, measurement, etc.) and the use of multidisciplinary teams, must be applied to support each of these activities to be effective. The full description of each OOSEM activity and process flows are provided in *A Practical Guide to SysML: Systems Modeling Language*[49] and in the *Object-Oriented Systems Engineering Method (OOSEM) Tutorial.*[50]

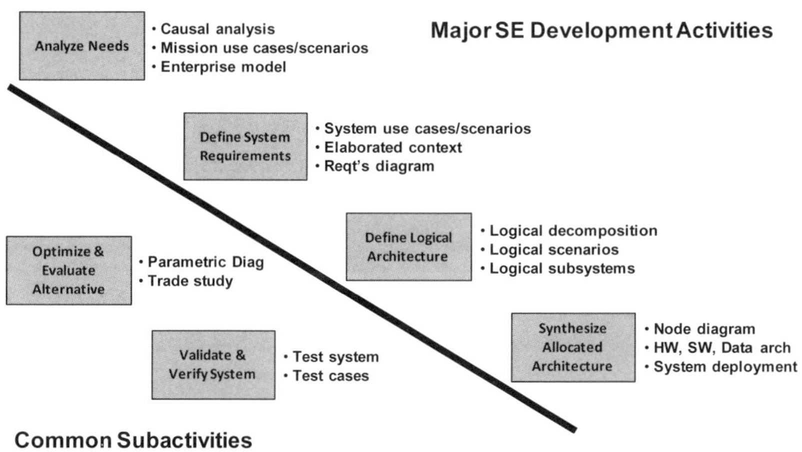

Figure 4-26 OOSEM Activities and Modeling Artifacts

4.12.3.2 Elaboration

Systems Modeling Language

Object-oriented SE utilizes a model-based approach to represent the various artifacts generated by the development activities using SysML™ as the predominant modeling language. As such, it enables the systems engineer to precisely capture, analyze, and specify the system and its elements and ensure consistency among various system views. The modeling artifacts can also be refined and reused in other applications to support product line and evolutionary development approaches.

OMG SysML™ is used to model complex systems and is an extension of the family of Unified Modeling Language (UML)-based standards that are intended to provide standard representations with well-defined semantics that can support model and data interchange. SysML™ has been developed as part of a joint initiative between OMG and INCOSE.[51,52,53]

SysML™ includes diagrams that can be used to specify system structure, parametric, requirement, and behavior and relationships. The modeling elements represented in the diagram facilitate integration among the various diagrams and views. The SysML™ diagram types shown in Figure 4-27 are summarized below.

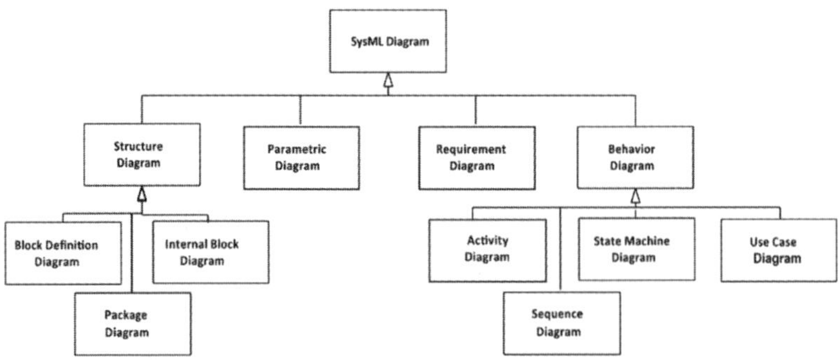

Figure 4-27 SysML™ Diagram Types

The system structure is represented as block definition diagrams and internal block diagrams. A block definition diagram describes the system hierarchy and system/system element classifications. The internal block diagram describes the internal structure of a system in terms of how its parts are inter-connected using ports and connectors. In addition, a package diagram is used to organize the model into packages that contain other model elements.

Behavior diagrams include the following:

- Use-case diagrams provide a high-level description of the system functionality in terms of how external systems use the system under consideration to achieve their goals.
- Activity diagrams represent the flow of data and control between activities.
- Sequence diagrams represent the interaction in terms of the time ordered exchange of messages between collaborating parts of a system.
- State machine diagrams describe the states of a system or its parts, and the transitions between the states in response to triggering events, along with the actions that occur upon transition, entry, exit of while in the state.

Requirements diagrams capture requirements hierarchies and the derivation, satisfaction, and verification relationships. It provides a bridge between requirements and system design models. The parametric diagram represents constraints on system parameter values, such as performance, reliability, and mass properties to support engineering analysis. SysML™ includes an allocation relationship to represent allocation of functions to elements, allocation of logical to physical elements, and other types of allocations.

The SysML™ diagrams are intended to capture the specification and design of a system. It is a general purpose modeling language that is intended to support many different MBSE methods, such as the one described in Section 4.12.1. Only a subset of the diagrams may be required to support a particular method. For example, a simplified functional analysis method may only require activity diagrams augmented by block definition diagrams, internal block diagrams, and perhaps requirements diagrams.

Additional information on SysML™ can be found at the official OMG SysML™ website at http://www.omgsysml.org. This site contains general information on SysML™, along with links to tool vendors, articles, and books.

Analyze Needs

This activity captures the "as-is" systems and organization, their limitations, and potential improvement areas. The results of the "as-is" analysis are used to develop the "to-be" organization and associated mission requirements. A Life-cycle model (see Section 7.1) depicts the organization, its constituent systems, including the systems to be developed or modified, and entities external to the organization. The "as-is" organization is analyzed using causal analysis techniques to determine its limitations, and used as a basis for deriving the mission requirements and "to-be" model. The mission requirements are specified in terms of the mission/organization needs, gaps, objectives, MOEs, and top-level use cases. The use cases and scenarios capture the mission and organization functionality.

Define System Requirements

This activity is intended to specify the system requirements that support the mission requirements. The system is modeled as a black box that interacts with the external systems and users represented in the Life-Cycle model. The system-level use cases and scenarios reflect the operational concept for how the system is used to support the organization. The scenarios are modeled using activity diagrams with swim lanes that represent the black box system, users, and external systems. The scenarios for each use case are used to derive the black box system functional, interface, data, and performance requirements. The requirements management database is updated during this activity to trace each system requirement to the organization/mission level use case and mission requirements.

Requirements variation is evaluated in terms of the probability that a requirement will change (e.g., sensitivity analysis), which is included in the risks, and later analyzed to determine how to design the system to accommodate the potential change. A typical example may be a system interface that is likely to change or a performance requirement that is expected to increase.

Define Logical Architecture

This activity includes decomposing and partitioning the system into logical elements that interact to satisfy the system requirements. The logical elements capture the system functionality. Examples may include a user interface that is realized by a web browser, or an environmental monitor that is realized by a particular sensor. The logical architecture/design mitigates the impact of requirements changes on the system design, and helps to manage technology changes.

Object-oriented SE provides guidelines for decomposing the system into its logical elements. The logical scenarios preserve system black box interactions with its environment. In addition, the logical system element functionality and data are repartitioned based on partitioning criteria, such as cohesion, coupling, design for change, reliability, performance, and other considerations.

Synthesize Allocated Architectures

The allocated architecture describes relationships among the physical system elements, including hardware, software, data, human, and procedures. The system nodes define the distribution of resources. Each logical element is first mapped to a system node to address how the functionality is distributed. Partitioning criteria is applied to address distribution concerns, such as performance, reliability, and security. The logical elements are then allocated to hardware, software, data, and manual procedure elements. The software, hardware, and data architectures are derived based on the element relationships. The requirements for each element are traced to the system requirements and maintained in the requirements management database.

Optimize and Evaluate Alternatives

This activity is invoked throughout all other OOSEM activities to optimize the candidate architectures and conduct trade studies to select the preferred architecture. Parametric models for modeling performance, reliability, availability, LCC, human, and other specialty engineering concerns are used to analyze and optimize the candidate architectures to the level needed to compare the alternatives. The criteria and weighting factors used to perform the trade studies are traceable to the system requirements and MOEs. This activity also includes the monitoring of TPMs and identifies potential risks. A critical aspect of architecture optimization centers on interfaces that require assessing alternatives based on interface concerns/optimization.

Validate and Verify System

This activity is intended to verify that the system design satisfies its requirements and to validate that the requirements meet the stakeholder needs. It includes the development of verification plans, procedures, and methods (e.g., inspection, demonstration, analysis, and test). System-level use cases, scenarios, and associated requirements are primary inputs to the development of the test cases and associated verification procedures. The verification system can be modeled using the same activities and artifacts described above for modeling the operational system. The requirements management database is updated during this activity to trace the system requirements and design information to the system verification methods, test cases, and results.

Tool Support

Tool support for OOSEM can be provided by COTS-based OMG SysML™ tools and associated requirements management tools. Other tools required to support the full system life-cycle should be integrated with the SysML™ and requirements management tools, such as configuration management, performance modeling, and verification tools. A more complete set of OOSEM tool requirements is provided in the referenced OOSEM tutorial.[49]

Offering/Availability

The OOSEM tutorial and training materials can be made available by contacting the INCOSE OOSEM Working Group to gain access through the INCOSE Connect collaboration space. Unlike other industry-provided Model-Based Systems Engineering (MBSE) methodologies, OOSEM is not a formal offering that can be purchased from any specific vendor, including professional services. However, support services may be available by contacting representatives of the INCOSE OOSEM Working Group.

4.13 References

1. ISO/IEC 15288:2008, p. 35
2. Duren, Riley. "Systems Engineering Challenges at NASA's Jet Propulsion Laboratory." *15th Annual INCOSE International Symposium*, 11 July 2005
3. Used with permission of Professor Heinz Stoewer
4. ISO/IEC 15288:2008, p. 36
5. Gilb, T., *Competitive Engineering,* Elsevier, 2005
6. Hook, I., *Customer-Centered Products: Creating Successful Products Through Smart Requirements Management*, Amacon, 2000
7. Adapted from Arnold, Stuart, and Harold Lawson, "Viewing Systems from a Business Management Perspective: The ISO/IEC 15288 Standard", *Journal of Systems Engineering,* 7:3, 2003
8. Heijden, K. vd, et al., *The Sixth Sense,* J. Wiley & Sons, 2002, p. 121
9. ISO/IEC 15288:2008, p. 39
10. ISO/IEC CD 29148, *Requirements Engineering* – currently under development
11. Roedler, Garry J., and Cheryl Jones, *Technical Measurement: A Collaborative Project of PSM, INCOSE, and Industry*, INCOSE-TP-2003-020-01, INCOSE Measurement Working Group, 27 December 2006
12. Martin, J. N., *Systems Engineering Guidebook*, CRC Press, 1996
13. see results of INCOSE vendor survey at <http://www.incose.org/ProductsPubs/products/toolsdatabase.aspx>
14. An example is Gilb's Planguage format, see <http://www.gilb.com>
15. EIA 632, *Standard – Processes for Engineering a System*
16. ISO/IEC 15288:2008, p. 40
17. see <http://www.computer.org/se vocab>
18. ISO 13407, *Human-centered design processes for interactive systems*
19. Wasson, Charles S., *System Analysis, Design, and Development*, NY: J. Wiley and Sons, 2006, p. 412
20. Zachman, John A., "A Framework for Information Systems Architecture," *IBM Systems Journal, vol. 26*, no. 3, 1987; IBM Publication G321-5298
21. The Zachman Institute for Framework Advancement (ZIFA) can be found at <http://www.zifa.com/>
22. Rechtin, Eberhardt and Mark W. Maier, *The Art of Systems Architecting*, 2nd Ed., Boca Raton, FL: CRC Press, 2000, p. 3
23. Christensen, Clayton M., *The Innovator's Dilemma*, NY: HarperCollins Publishers, 2000
24. Adams, James L, (1990). *Conceptual Blockbusting*, 3rd Ed., San Francisco Book Company, Inc.
25. ISO/IEC 15288:2008, p. 42
26. Gilb, T., and Dorothy Graham, *Software Inspection*, Addison-Wesley Longman, 1993

27. ISO/IEC 15288:2008, p. 44
28. Ibid, p. 45
29. *Systems Verification, Validation and Testing Methodology Guidelines*, Contract: G1RD-CT-2002-00683, <http://www.incose.org/secoe/0105.htm>
30. Ibid
31. Picture downloaded from <http://www.jk-labs.com/ProdDev.html>, May 22, 2006.
32. ISO/IEC 15288:2008, p. 46
33. Ibid, p. 47
34. Ibid, p. 49
35. Ibid, p. 50
36. Ibid, p. 52
37. see <http://www.zerofootprint.net/ and www.zeri.org/>
38. see <http://www.iso-14001.org.uk/>
39. see <http://www.mcdonough.com/>
40. Grady, Jeffrey O., "Universal Architecture Description Framework." *INCOSE Journal of Systems Engineering: The Journal of the International Council on Systems Engineering*. Vol. 12, Issue 2 (Summer 2009), 91-110
41. Wymore, A. W., *Model-based Systems Engineering*, CRC Press, 1993
42. Friedman, G., *Constraint Theory: Multidimensional Mathematical Model Management*, Springer, 2005
43. Becker, Ofri, Joseph Ben-Ashe, and Ilya Ackerman, "A Method for System Interface Reduction Using N^2 Charts," *Systems Engineering, The Journal of the International Council on Systems Engineering*, Volume 3, Number 1, Wiley, 2000
44. Defense Systems Management College, "Systems Engineering Management Guide," Fort Belvoir, VA, 3 Oct 83, page 6-5
45. Lano, R. *The N^2 Chart.* TRW Inc., 1977
46. An extract from Estefan, Jeff. "Survey of Model-Based Systems Engineering (MBSE) Methodologies," Rev B, Section 3.2. NASA Jet Propulsion Laboratory, June 10, 2008. The document was originally authored as an internal JPL report, and then modified for public release and submitted to INCOSE to support the INCOSE MBSE Initiative.
47. Lykins, Howard, Sanford Friedenthal, and Abraham Meilich, "Adapting UML for an Object-Oriented Systems Engineering Method (OOSEM)," *Proceedings of the INCOSE 2000 International Symposium*, Minneapolis, MN, July 2000
48. Friedenthal, Sanford, "Object Oriented Systems Engineering," *Process Integration for 2000 and Beyond: Systems Engineering and Software Symposium*. New Orleans, LA, Lockheed Martin Corporation, 1998
49. Friedenthal, Sanford, Alan Moore, and Rick Steiner, "Chapter 16," *A Practical Guide to SysML: Systems Modeling Language*, Morgan Kaufmann Publishers, Inc., July 2008
50. "Object-Oriented Systems Engineering Method (OOSEM) Tutorial," Ver. 03.00, Lockheed Martin Corporation and INCOSE OOSEM Working Group, October 2008
51. Friedenthal, Sanford, and Roger Burkhart, "Extending UML from Software to Systems," *13th Annual INCOSE International Symposium*. Arlington, VA, June 30- July 1, 2003.
52. Friedenthal, Sanford, and Cris Kobryn, "Extending UML to Support a Systems Modeling Language", *14th Annual INCOSE International Symposium*, Toulouse, France, June 2004
53. OMG Systems Engineering Domain Special Interest Group (SE DSIG) Website, <http://syseng.omg.org>

This page intentionally left blank.

5 Project Processes

Within the system life cycle, the creation or upgrade of products and services is managed by the conduct of projects. For this reason, it is important to understand the contribution of SE to the management of the project. Project Processes are defined in ISO/IEC 15288:2008 as follows:

> *The Project Processes are used to establish and evolve project plans, to execute the project plans, to assess actual achievement and progress against the plans and to control execution of the project through to fulfillment. Individual Project Processes may be invoked at any time in the life cycle and at any level in a hierarchy of projects, as required by project plans or unforeseen events. The Project Processes are applied with a level of rigour and formality that depends on the risk and complexity of the project.*[1]

Systems engineers continually interact with project management. One example of this interaction is illustrated in Figure 5-1 below.

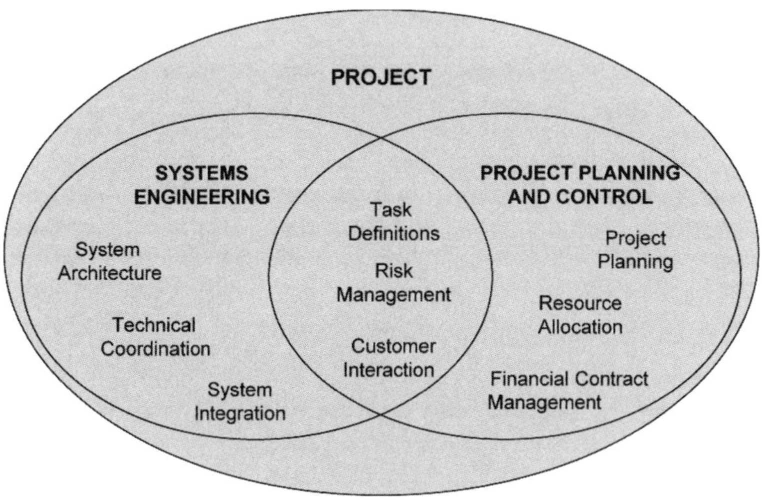

Figure 5-1 SE/Project Planning and Control Overlap[2]

Project Processes include: Project Planning, Project Assessment and Control, Decision Management, Risk Management, Configuration Management, Information Management, and Measurement. These processes are found throughout an organization as they are essential to generic management practices and apply both inside and outside the project context. This chapter of the handbook focuses on processes relevant to the technical coordination of a project. Table 5-1 contains the list of acronyms that appear in the context diagrams in this chapter.

Table 5-1 Acronym List

CMP	Configuration Management Plan
IMP	Information Management Plan
QMP	Quality Management Plan
RMP	Risk Management Plan
SEP	Systems Engineering Plan
WBS	Work Breakdown Structure

5.1 Project Planning Process

5.1.1 Overview

5.1.1.1 Purpose

As stated in ISO/IEC 15288:2008:

> *The purpose of the Project Planning Process is to produce and communicate effective and workable project plans.*
>
> *This process determines the scope of the project management and technical activities, identifies process outputs, project tasks and deliverables, establishes schedules for project task conduct, including achievement criteria, and required resources to accomplish project tasks.*[3]

Project planning establishes the direction and infrastructure necessary to assess and control the progress of a project and identifies the details of the work and the right set of personnel, skills, and facilities with a schedule of need for resources from within and outside the organization.

5.1.1.2 Description

Project planning starts with a statement of need, often expressed in a project proposal. The Project Planning Process is performed in the context of the organization. Life Cycle Model Management Processes (see Section 7.1) establish and identify relevant policies and procedures for managing and executing a technical effort; identifying the technical tasks, their interdependencies, risks, and opportunities; and providing estimates of needed resources and budgets. This is also the point in the process to determine the need for specialized equipment or facilities and specialists during the project life cycle to improve efficiency and effectiveness and decrease cost over-runs.

For example, during product design, various disciplines work together to evaluate parameters such as manufacturability, testability, operability, and sustainability against product performance. In some cases, project tasking is concurrent to achieve the best results. Figure 5-2 shows the Project Planning Process Context Diagram.

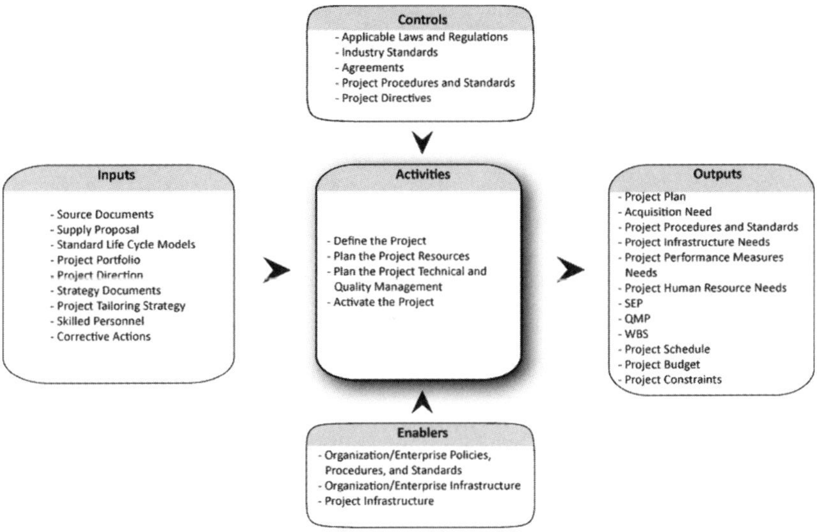

Figure 5-2 Context Diagram for the Project Planning Process

5.1.1.3 Inputs

The inputs to the Project Planning Process include the following:

- *Source Documents*
- *Supply Proposal* – Proposal and technical results from the initial concept exploration stage
- *Standard Life-cycle Models* – Need to select the model or models appropriate for the project
- *Project Portfolio* – Including authorization to initiate the project and project goals
- *Project Direction* – Organizational direction to the project, includes sustainment of projects meeting assessment criteria and redirection or termination of projects not meeting assessment criteria
- *Strategy Documents* – Including:
  - Organization Strategic Plan

  - Implementation Strategy
  - Integration Strategy
  - Verification Strategy
  - Transition Strategy
  - Validation Strategy
  - Operation Strategy
  - Maintenance Strategy
  - Disposal Strategy
  - Decision Management Strategy
  - Risk Strategy
  - Configuration Management Strategy
  - Information Management Strategy
  - Measurement Strategy
  - Acquisition Strategy
  - Supply Strategy
- *Project Tailoring Strategy* – Describes the project-specific tailoring of the organization's standard processes
- *Skilled Personnel* – The right people with the right skills at the right time
- *Project Directives* – Internal projects directives based on assessment and control activities
- *Corrective Actions* – Resulting from project-related reviews and audits.

This process is governed by the following controls and enablers:

- *Applicable Laws and Regulations*
- *Industry Standards* – relevant industry specifications and standards
- *Agreements* – terms and conditions of the agreements
- *Project Procedures and Standards* – including project plans
- *Project Directives*
- *Organization/Enterprise Policies, Procedures, and Standards* – including guidelines and reporting mechanisms

- *Organization/Enterprise Infrastructure*
- *Project Infrastructure.*

5.1.1.4 Outputs

Outputs of the Project Planning Process include the following:

- *Project Plan* – May be a single plan; on larger projects, may consist of multiple plans.
- *Acquisition Need* – For systems, subsystems, and system elements that need to be acquired.
- *Project Procedures and Standards* – Project-unique procedures and standards to guide the technical effort.
- *Project Infrastructure Needs* – Infrastructure needs, including commitments to external stakeholders, are derived from and require coordination with the organization.
- *Project Performance Measures Needs* – Identification of measures that are analyzed to identify any deviations from the plans or technical performance of the product, then used to make decisions regarding future work and technical options.
- *Project Human Resource Needs* – Personnel needs are derived from and require coordination with the organization.
- *Systems Engineering Plan (SEP)* – Includes identification of required technical reviews and their completion criteria, methods for controlling changes, risk and opportunity assessment and methodology, and the identification of other technical plans and documentation to be produced for the project.
- *Quality Management Plan (QMP)*
- *Work Breakdown Structure (WBS)* – Including a data dictionary.
- *Project Schedule* – May include a top-level milestone schedule and multiple levels (also called tiers) of schedules of increasing detail and task descriptions with completion criteria and work authorizations.
- *Project Budget* – Includes labor, infrastructure, acquisition, and enabling system costs along with reserves for risk management.
- *Project Constraints* – Anything that may limit or restrict the project or system solution.

5.1.1.5 Process Activities

The Project Planning Process includes the following activities:

- *Define the Project*
  - Analyze the project proposal and related agreements to define the project scope
  - Identify project objectives and project constraints
  - Establish tailoring of organization procedures and practices to carry out planned effort. Define and maintain a life cycle model that is tailored from the defined life cycle models of the organization. Chapter 8 contains a detailed discussion on tailoring.
  - Establish tailoring of organization procedures and practices to carry out planned effort. Chapter 8 contains a detailed discussion on tailoring
- *Plan Project Resources*
  - Establish the roles and responsibilities for project authority
  - Define top-level work packages for each task and activity identified. Each work package should be tied to required resources including procurement strategies.
  - Develop a project schedule based on objectives and work estimates
  - Define the infrastructure and services required
  - Define costs and estimate project budget
  - Plan the acquisition of materials, goods and enabling system services
- *Plan Project Technical and Quality Management*
  - Prepare an SEP; tailor the Quality, Configuration, Risk and Information Management plans to meet the needs of the project
  - Tailor the organization Risk Management Processes and practices in accordance with the agreements and the SEP to establish a systematic approach for identifying and handling risk
  - Tailor the organization Configuration Management Processes and practices in accordance with the agreements and the SEP to

establish a systematic approach for identifying and handling change requests

- *Activate the Project.*

Common approaches and tips:

- An important outcome of planning is the SEP. This plan identifies the activities to be accomplished, key events that must be satisfied at decision gates throughout the project, work packages that define the working schedule, and the assignment of required resources (i.e., people, equipment, and facilities) that define the project budget. Each decision gate will have a list of activities or tasks that must be successfully completed prior to entering the decision gate. This plan references other planning instruments that are tailored for use on the project, such as the Configuration Management, Risk Management, and Information Management Plans discussed in later sections of this chapter.
- IPDTs are frequently used to break down communications and knowledge stovepipes within organizations.[4]
- Creation of the WBS is an activity where SE and project management intersect.[5]
- Skipping or taking shortcuts in the planning process reduces the effectiveness of other Project Processes.
- Even agile project management methods include planning – the cycles may be shorter and more frequent, but planning is an essential process.
- Incorporate risk assessment early in the planning process to identify areas that need special attention or contingencies. Always attend to the technical risks.
- The Project Management Institute is a source of guidelines for project planning.[6]
- The ISO/IEC 16326 standard for project management also provides additional guidance on this subject.

5.1.2 Elaboration

5.1.2.1 Project Planning Concepts

Project Planning estimates the project budget and schedule against which project progress will be assessed and controlled. Systems engineers and project managers must collaborate in project planning. Systems engineers perform technical management activities consistent with project objectives. Technical management activities include planning, scheduling, reviewing, and auditing the SE process as defined in the SEP, and the SE Master Schedule (SEMS).

5.1.2.2 Systems Engineering Plan

The SEP (also called Systems Engineering Management Plan, or SEMP) is the top-level plan for managing the SE effort and, as such, defines how the project will be organized, structured, and conducted and how the total engineering process will be controlled to provide a product that satisfies stakeholder requirements. A well-written SEP provides guidance to a project and helps the organization avoid unnecessary discussions about how to perform SE. Organizations generally maintain a template of the SEP suitable for tailoring and reuse. Effective project control requires that there be a SEP, which the systems engineer keeps current and uses on a daily basis to manage the team's actions.

> Note: The U.S. DOD and others use the SEP to describe how the government will manage their SE efforts. They use the term SEMP to describe how the contractor will manage their SE efforts. This handbook uses SEP for both.

The SEMS is an essential part of the SEP and a tool for project control because it identifies the critical path of technical activities in the project. Verification activities may also receive special attention in the SEMS. In addition, the schedule of tasks and dependencies helps justify requests for personnel and resources needed throughout the development lifecycle.

The SEP and SEMS are supported by a project or contract WBS that defines a project task hierarchy. Work authorization is the process by which the project is baselined and financially controlled. A description of the organization procedures for starting work on a part of the WBS may be defined in the SEP.

Technical Performance Measures (see Section 5.7.2.4) are a tool used for project control, and the extent to which TPMs will be employed should be defined in the SEP.[7]

A SEP should be prepared early in the project, submitted to the customer (or to management for in-house projects), and used in technical management for the concept and Development Stages of the project, or the equivalent in commercial practice. Creation of the SEP involves defining the SE processes, functional analysis approaches, what trade studies will be included in the project, schedule, and organizational roles and responsibilities, to name a few of the more important aspects of the plan. The SEP also reports the results of the effort undertaken to form a project team and outlines the major deliverables of the project, including a decision database, specifications, and baselines. Participants in the creation of the SEP should include senior systems engineers, representative subject matter experts, project management, and often the customer.

The format of the SEP can be tailored to fit project, customer, or company standards. Usually, SE organizations maintain a focus on SE processes. To maximize reuse of the SEP for multiple projects, project-specific appendices are often used to capture detailed and dynamic information, such as the decision database, a schedule of milestones and decision gate reviews, and the methodology to be used in resolving problems uncovered in reviews.

The process inputs portion of the SEP identifies the applicable source documents (e.g., customer specifications from the RFP, SOW, industry standards, etc.) to be used in the performance of the project and in developing associated deliverables (e.g., the Systems Specification and Technical Requirements Document). It also may include previously developed specifications for similar systems and company procedures affecting performance specifications. A technical objectives document should be developed and may be one of the source documents for the decision database. The document may also be part of the ConOps for the system (see Section 4.1.2.5).

The approach and methods used to define the performance and functional requirements for the following SE and Specialty Engineering areas should also be documented:

1. Organization of the project and how SE interfaces with the other parts of the organization. How are communications at these interfaces handled? How are questions and problems elevated up the organization and resolved?
2. Responsibilities and authority of the key positions
3. Clear system boundaries and scope of the project
4. Project assumptions and constraints
5. Key technical objectives
6. Risk and opportunity plan, assessment, and methodology
7. Validation planning (not just verification planning)
8. Configuration Management planning
9. QA planning

10. Infrastructure support and resource management (i.e., facilities, tools, IT, personnel, etc.)
11. Reliability, availability, maintainability, supportability, and Integrated Logistics Support (ILS)
12. Survivability, including nuclear, biological, and chemical
13. EMC, radio frequency management, and electrostatic discharge
14. Human Engineering and HSI
15. Safety, health hazards, and environmental impact
16. System security
17. Producibility
18. Test and evaluation
19. Testability and integrated diagnostics
20. Computer resources
21. Transportability
22. Other engineering specialties bearing on the determination of performance and functional requirements.

Under some circumstances, the SEP may address Design-to-Cost (see Section 9.6) and Value Engineering (see Section 9.13) practices to provide insight into system/cost-effectiveness. For example, can the project be engineered to have significantly more value with minimal additional cost? If so, does the customer have the resources for even the modest cost increase for the improvement? The intent is to assure the customer that no obvious cost-effective alternatives have been overlooked.

Technical reviews are essential to ensure that the system being developed will meet requirements and that the requirements are understood by the development team. Formal reviews are essential to determine readiness to proceed to the next stage of the system life cycle. The number and frequency of these reviews and their associated decision gates must be tailored for specific projects. The SEP should list what technical reviews will be conducted and the methodology to be used in solving problems uncovered during those reviews.

The system life cycles shown in Figure 3-3 illustrate the appropriate time for reviews and decision gates. They may or may not be right for a given project, and some projects may need more or fewer reviews. Additionally, formal, documented decision gates, with the customer in attendance, can impose significant cost on the project. As such, projects should plan to use more-frequent, informal, in-house reviews to resolve most issues; and strive to exit decision gates with no major customer-imposed Action Items.

Transitioning critical technologies should be done as a part of Risk Management (see Section 5.4), but is called out separately here for special emphasis. Critical technologies should be identified and the steps outlined for risk management should be followed. Additionally, completed and planned Risk Management work should be explicitly referenced in the SEP.

The system being proposed may be complex enough that the customer will require training to use it. During the project, it may be necessary to train those who will develop, manufacture, verify, deploy, operate, support, conduct training, or dispose of the system. A plan for this training is required in the SEP and should include the following:

1. Analysis of performance
2. Behavior deficiencies or shortfalls
3. Required training to remedy deficiencies or shortfalls
4. Schedules to achieve required proficiencies.

Verification is usually planned using a verification matrix that lists all the requirements and anticipated verification methods. The possible methods of verification include inspection, analysis, demonstration, and test. The SEP should state, at least in preliminary general terms, that a verification plan will be written to define the items to be verified and which methods will be used to verify performance. The plan should also define who is to perform and witness the verification of each item. This should also relate to the SEMS for time phasing of the Verification Process. Detailed procedures are usually not written for inspection, analysis, and demonstration methods. Simulations may be used for testing, when quantifiable results are needed, or for demonstration, when qualitative results are satisfactory.

A well-written SEP provides guidance to a project and helps the organization avoid unnecessary discussions about how to perform SE. In addition, a schedule and organization are defined that help the project procure the personnel needed throughout the development lifecycle and assess progress.

5.1.2.3 *Integrated Product and Process Development*

Integrated Product Development (IPD) evolved from recognizing the need to consider all elements of the product life cycle, from conception through disposal, starting at the beginning of the life cycle. Important items to be considered include quality, cost, schedule, user requirements, manufacturing, and support. Further, IPD implies the continuous integration of the entire product team, including engineering, manufacturing, verification, and support, throughout the product life cycle.

Historically, traditional development took place in series, with one activity starting as the preceding one was completed. This is a very lengthy process, and the product could become obsolete before it is completed. With good interface definition and control, IPD, involving the entire team, can speed up the development process.

In the early 1990s, companies began to discover that they really could be more productive and reduce the risks inherent in concurrent product development if they moved away from the traditional hierarchical management structure and organized into Integrated Product Teams (IPTs). Some of the greatest productivity gains came in three areas:

- Unleashing the team's ingenuity through decentralized processes
- Avoidance of previous problems through new, creative approaches
- Better integration between engineering and manufacturing.

In turn, these led to improved product features, performance, quality, and customer satisfaction. Later, as the importance of *process* was recognized, the terminology was modified to *Integrated Product & Process Development,* or IPPD. The following definitions apply to this concept:

- *Integrated Product Development Team (IPDT)* – A multidisciplinary group of people who are collectively responsible for delivering a defined product or process.
- *Integrated Product & Process Development (IPPD)* – The process of using IPDTs to simultaneously develop the design for a product or system and the methods for manufacturing the product or system. The process verification may consist of review of a process description by an IPDT. It may also include a demonstration to an IPDT of a process.
- *Concurrent Engineering* – Is a management/operational approach which aims to improve product design, production, operation, and maintenance by developing environments in which personnel from all disciplines (i.e., design, marketing, production engineering, process planning, and support) work together and share data throughout all stages of the product life cycle.

Integrated development has the potential to introduce *more* risk into a development program because downstream activities are initiated on the *assumption* that upstream activities will meet their design and interface requirements. However, the introduction of a hierarchy of cross-functional IPDTs, each developing and delivering a product, has been found to actually reduce risks and provide better products faster. Hence, the objectives of using IPPD are as follows:

- Reduce time to market
- Improve product quality
- Reduce waste
- Save costs through the complete integration of SE life-cycle processes.

In addition to these tangible outcomes, IPPD improves team communications through IPDTs, implements a proactive risk process, makes decisions based on timely input from the IPDT, and improves customer involvement.

<u>Integrated Product Development Team Overview</u>

An IPDT is a process-oriented, integrated set of cross-functional teams (i.e., an overall team comprised of many smaller teams) given the appropriate resources and charged with the responsibility and authority to define, develop, produce, and support a product or process (and/or service). Process orientation means that each team is staffed with all the skills necessary to complete their assigned processes, which may include all or some of the development and production steps.

As noted above, industry has learned that IPDTs, using best practices and continuous improvement, achieve significant *process* improvements resulting in:

- Seamless interfaces within the teams
- Reduced engineering design time
- Fewer problems in transition from engineering to manufacturing
- Reduced development time and cost.

The general approach is to form cross-functional IPDTs for all products and services. There are typically three types of IPDTs:

1. Systems Engineering and Integration Team (SEIT)
2. Product Integration Team (PIT)

3. Product Development Team (PDT).

These integrated teams each mimic a small, independent project focusing on individual elements and/or their integration into more-complex subsystems and elements. The SEIT balances requirements between product teams, helps integrate the other IPDTs, focuses on the integrated system and system processes, and addresses systems issues, which, by their nature, the other IPDTs would most likely relegate to a lower priority. Although the teams are organized on a process basis, the organizational structure of the team-of-teams may approach a hierarchical structure for the product, depending upon the way the product is assembled and integrated.

The focus areas for the three types of IPDT teams and their general responsibilities are summarized in Table 5-2. This arrangement is often applicable to large, multi-element, multiple subsystem programs, but must be adapted to the specific project. For example, on smaller programs, the number of PIT teams can be reduced or eliminated. In service-oriented projects, the system hierarchy, focus, and responsibilities of the teams must be adapted to the appropriate services.

Team members' participation will vary throughout the product cycle, and different members may have primary, secondary, or minor support roles as the effort transitions from requirements development through the different stages of the project life cycle. For example, the manufacturing and verification representatives may have minor, part-time advisory roles during the early product definition stage, but will assume primary roles later, during manufacture and verification. The idea is to have team members participate to the degree necessary from the outset to ensure their needs and requirements are reflected in overall project requirements and planning to avoid costly changes later. Further, it is good for at least some of the team to remain throughout the product cycle to retain the team's "project memory."

Finally, IPDTs must be empowered with life-cycle (Concept-to-Retirement) responsibility for their products and systems and with the authority to get the job done. Further, they should <u>not</u> be looking to higher management for key decisions. They should, however, be required to justify their actions and decisions to others, including interfacing teams, the system integration team, and project management.

Table 5-2 Types of IPDTs, their Focus and Responsibilities

System Hierarchy	Team Type + Focus Responsibilities
External Interface & System	System Engineering & Integration Team (SEIT)
	• Integrated System and Processes • External & Program Issues • System Issues & Integrity • Integration & Audits of Teams
Upper-level Elements	Product Integration Teams (PITs)
	• Integrated H/W and S/W • Deliverable Item Issues & Integrity • Support to Other Teams (SEIT and PDTs)
Lower-level Elements (e.g., Assemblies, Components, and Parts)	Product Development Teams (PDTs)
	• Hardware and Software • Product Issues & Integrity • Primary Participants (Design and Mfg.) • Support to Other Teams (SEIT and PITs)

Integrated Product Development Team Process

As stated above, the basic principle of IPDT is to get all disciplines involved at the beginning of the product development process to ensure that needs and requirements are completely understood for the full life cycle of the product. Historically, the initial development of requirements has been led by systems engineers. In an IPDT, systems engineers still lead the requirements development process, but now more (preferably all) disciplines participate.

Requirements are developed initially at the system level, then successively at lower levels as the requirements are flowed down. Teams, led by systems engineers, perform the up-front SE activities at each level. This is *different* from the previous, classical development approach where systems engineers did the up-front work and passed the requirements along to development engineers who passed their designs on to manufacturing, and so forth, without the continuous involvement of the initial engineers. Ultimately, this resulted in a loss of understanding caused by asynchronous communications.

In an IPPD environment, IPDTs do their own internal integration. A SEIT representative belongs to each product team (perhaps several); with both internal and external team responsibilities. There is extensive iteration between the product teams and the SEIT to converge on requirements and design concepts, although this effort should slow down appreciably after the preliminary design review and as the design firms up.

Systems engineers participate heavily in the SEIT and PIT and to a much lesser extent in the PDT. Regardless, the iterative SE processes described in this handbook are just as applicable to all teams in the IPPD environment as they were in previous styles of organization. In fact, it is easier to apply the processes throughout the program because of the day-to-day presence of systems engineers on all teams.

All IPDTs have many roles, and their integration roles overlap, based on the type of team and the integration level. Figure 5-3 gives examples for various program processes and system activities. In this figure, Program Processes covers just about anything required on the program. The three bars on the left of the figure show the roles of the three types of product teams at different levels of the system. Note, for example, that the SEIT leads and audits in external integration and in system integration activities, as indicated by the shaded bar. But, for those program processes involving lower-level elements (e.g., parts, components, or subassemblies), the appropriate PDTs are the active lead and audit participants, supported by the SEIT and the PIT.

Basic system activities include system requirements derivation, system functional analysis, requirements allocation and flowdown, system trade-off analysis, system synthesis, system integration, TPM, and system verification. The bars for activities 1, 2, and 3 in the chart show that the SEIT leads and audits activities on different system activities while the element teams participate. The lower-level system element (e.g., assembly, subassembly, component, and part) teams provide additional support, if requested.

The column at the right side of Figure 5-3 shows other integration areas where all teams have some involvement. The roles of the various teams must also be coordinated for these activities, but should be similar to the example.

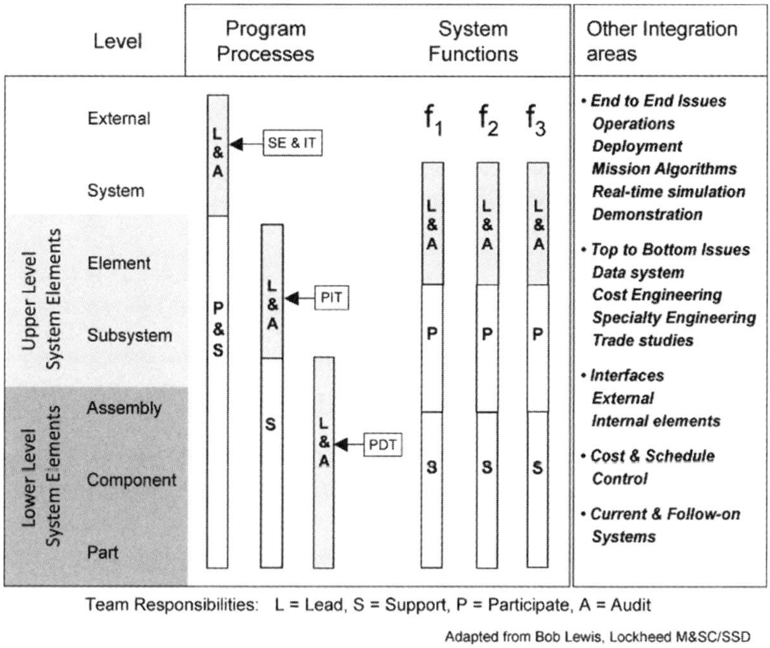

Figure 5-3 Examples of Complementary Integration Activities of IPDTs

Organizing and Running a High-Performance IPDT

The basic steps necessary to organize and run an IPDT on a project are as follows. Each step is discussed in turn, with a summary of the key activities that should take place during the step.

1. *Define the IPDT teams for the project* – Develop IPDT teams that cover all project areas.
2. *Delegate responsibility and authority to IPDT leaders* – Select experienced team leaders early in the development process and avoid frequent budget changes through-out the life cycle.
3. *Staff the IPDT* – Candidates must work well in a team environment, communicate well, and meet their commitments.
    - Balance the competency, availability, and full-time commitment of the core team.
    - Plan when competencies are needed and not needed.
    - Identify issues where specialists are needed.

4. *Understand the team's operating environment* – Recognize how the team directly or indirectly influences other teams and the project as a whole.

5. *Plan and conduct the "Kick-Off Meeting"* – Recommend two kick-off meetings, one for the project as a whole and one for the individual IPDTs. Well planned kick-off meetings will set the project off on the right foot.

6. *Train the team* – Training for the project is a critical element. The following recommended topics should be covered:

    - Tailored SE process for the project
    - Project description, customer stakeholders, purpose, mission, organization, schedule, budget
    - Terminology and nomenclature
    - Access to project products
    - Communications skills
    - Team-of-Teams concept
    - Project IPDT procedures, measures, and reporting.

    Additional training sessions should be held and self learning guides should be developed to help new team members come up to speed on the project when staff turnover occurs.

7. *Define the team vision and objectives* – Use collaborative brainstorming in the initial IPDT meetings to develop the team's vision and objectives such that each member has an ownership. It most likely will be necessary to bring in other IPDT members, management, and customers to flesh out the vision and objectives of the team.

8. *Have each team expand the definition of its job* – Once the higher level project plan has been reviewed, each team must identify the tasks, roles, responsibilities, and milestones of the team and each of the members. Members need to understand how their individual tasks fit into the higher level project-program tasks.

9. *Establish an expectation of routine Process Assessment and Continuous Improvement* – Each team must document the process they are using and the key measures to be monitored. The teams must have the mindset of continuous improvement, monitor their own activities, and continually make course corrections along the way.

10. *Monitor team progress via measures and reports* – Each team will have a set of measures and reports to monitor its own progress. These reports and measures must be reviewed by the SEIT that coordinates among the other IPDTs. These measures may include an Earned Value Report and technical measures, such as a Defect Rate report. The selected measures are dependent on the team's role on the project.

11. *Sustain and evolve the team throughout the project* – Personnel assignments to a team will vary as each team grows, shrinks, and changes skill mix over the project life cycle. As issues arise, technical specialists may need to join the team to help address these specific issues. Services such as marketing, program controls, procurement, finance, legal, and human resources generally support the team at a steady, low level of effort, or as required.

12. *Document team products* – The team's products should be well defined and do not usually change significantly. Because of the IPDT structure, the overhead of cross-organizational communication varies. Ideally, this would be reduced or eliminated. When multiple documents are required, different team members, with identified backups, should be assigned as the responsible author with contributions from other team members. The IPDT should maintain a log of activities in addition to the mission, vision, objectives, deliverables, meeting minutes, decisions, tailored processes, agreements, team project information, and contact information.

13. *Close the project and conduct follow-up activities* – In conjunction with step 12, the IPDT should maintain records as though the project may be re-engineered at some future time and all close-out products must be accessible. All IPDT logs should be organized the same way, when possible, such that they can be easily integrated into an overall project report. The close-out should include lessons learned, recommended changes, and a summary of measures for the team.

Project managers should review team staffing plans to ensure proper composition and strive for continuity of assignments. It has been observed that the advantages of a full-time contributor outweigh the work of many part-time team members. Similarly, the loss of a key team member who knows how and why things are done can leave the team floundering. On IPDTs, it is important to have people who can work well together and communicate. However, team results may be condemned to mediocrity by avoiding outstanding technical specialists and professionals who can really make a difference. Table 5-3 lists ten techniques for achieving high performance in an IPDT.

Table 5-3 Ten Techniques for High Performance in Integrated Product Development Teams (IPDTs)

	Recommended Technique
1.	Carefully select the staff – Excellent people do excellent work.
2.	Establish and maintain positive team interaction dynamics – all should know what is expected of the team and each individual, all should strive to meet commitments, interactions should be informal but efficient, and a "no blame" environment where problems are fixed and the team moves on
3.	Generate team commitment and buy-in –to the vision, objectives, tasks, and schedules
4.	Breakdown the job into manageable activities – those that can be accurately scheduled, assigned, and followed-up on weekly
5.	Delegate and spread out routine administrative tasks among the team – frees the leader to participate in technical activities, give every team member some administrative/ managerial experience
6.	Create a "world class" analysis and simulation capability – for requirements and performance to be better than the competition
7.	Schedule frequent team meetings with mandatory attendance for quick information exchanges – everyone is current; assign action items with assignee and due date
8.	Maintain a Team Leader's Notebook
9.	Anticipate and surface potential problems quickly (internally and externally)
10.	Acknowledge and reward good work.

Potential IPDT Pitfalls

Ample opportunities exist to go astray before team members and leaders go through several project cycles in the IPDT framework and gain the experience of working together. As such, there are some things teams should watch out for. Table 5-4 describes eight pitfalls common to the IPDT environment.

Table 5-4 Pitfalls of using IPDT

	IPDT Pitfalls	What to do
1.	Spending too much time defining the vision and objectives	Converge and move on
2.	Insufficient authority – IPDT members must frequently check with management for approval	Give team leader adequate responsibility, or put the manager on the team
3.	IPDT members are insensitive to management issues and over commit or overspend	Team leader must remain aware of overall project objectives and communicate to team members

	IPDT Pitfalls	What to do
4.	Teams are functionally-oriented rather than cross-functionally process-oriented	Review step 1 Organizing and Running an IPDT (see above)
5.	Insufficient continuity of team members throughout the project	Management should review staffing requirements
6.	Transition to the next stage team specialists occurs too early or too late in the schedule	Review staffing requirements
7.	Overlapping assignments for support personnel compromises their effectiveness	Reduce the number of teams
8.	Inadequate project infrastructure	Management involvement to resolve

As noted in Pitfall 2, some things do require checking with higher authority. Encourage team members to anticipate these from the outset. Functional managers and supervisors, if any, must remain aware of major team issues and coach, guide, and train participants until they gain the requisite experience.

5.2 Project Assessment and Control Process

5.2.1 Overview

5.2.1.1 Purpose

As stated in ISO/IEC 15288:2008:

> *The purpose of the Project Assessment and Control Process is to determine the status of the project and direct project plan execution to ensure that the project performs according to plans and schedules, within projected budgets, to satisfy technical objectives.*
>
> *This process evaluates, periodically and at major events, the progress and achievements against requirements, plans and overall business objectives. Information is communicated for management action when significant variances are detected. This process also includes redirecting the project activities and tasks, as appropriate, to correct identified deviations and variations from other project management or Technical Processes. Redirection may include re-planning as appropriate.*[8]

Assessments are scheduled periodically and for all milestones and decision gates. The intention is to maintain good communications within the project team and with the stakeholders, especially when deviations are encountered.

The process uses these assessments to direct the efforts of the project, including redirecting the project when the project does not reflect the anticipated maturity.

5.2.1.2 Description

The Project Planning Process (see Section 5.1) identified details of the work effort and expected results. The Project Assessment and Control Process collects data to evaluate the adequacy of the project infrastructure, the availability of necessary resources, and compliance with project performance measures. Assessments also monitor the technical progress of the project, and may identify new risks or areas that require additional investigation. A discussion of the creation and assessment of TPMs is found in Section 5.7.2.4.

The rigor of the Project Assessment and Control Process is directly dependent on the complexity of the system-of-interest. Project control involves both corrective and preventive actions taken to ensure that the project is performing according to plans and schedules and within projected budgets. The Project Assessment and Control Process may trigger activities within the other process areas in this chapter. Figure 5-4 is the context diagram for the Project Assessment and Control Process.

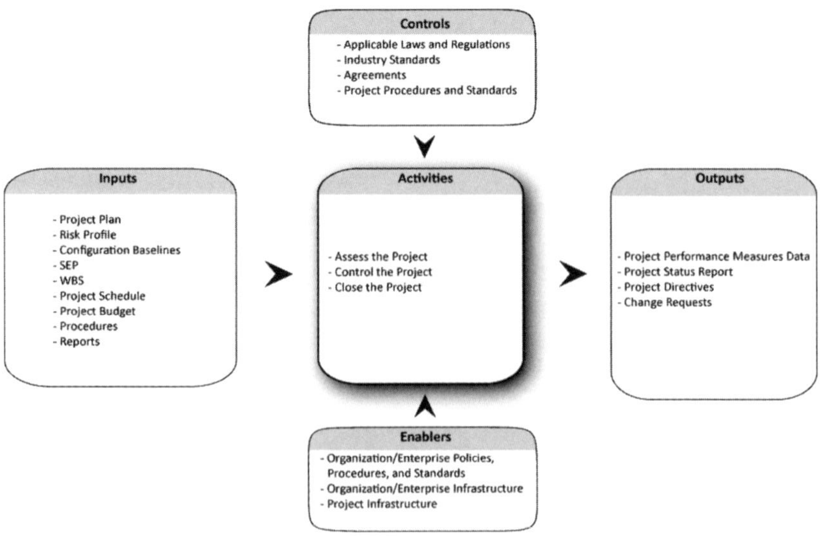

Figure 5-4 Context Diagram for the Project Assessment and Control Process

5.2.1.3 Inputs

Inputs to the Project Assessment and Control Process include:

- The project baseline documented during the Project Planning Process:
  - *Project Plan*
  - *Systems Engineering Plan (SEP)*
  - *WBS*
  - *Project Schedule*
  - *Project Budget*
- *Procedures* – Procedures generated and used in the life cycle processes including:
  - Integration Procedure
  - Verification Procedure
  - Transition Procedure
  - Validation Procedure
  - Maintenance Procedure
  - Disposal Procedure
- *Reports* – Reports generated from the life cycle processes including:
  - Integration Report
  - Verification Report
  - Transition Report
  - Validation Report
  - Operation Report
  - Maintenance Report
  - Disposal Report
  - Decision Report
  - Risk Report
  - Configuration Management Report
  - Information Management Report
  - Measurement Report

  - Acquisition Report
  - Supply Report
  - Infrastructure Management Report
- *Other inputs* – including:
  - Risk Profile (sometimes referred to as a risk matrix)
  - Configuration Baselines.

This process is governed by the following controls and enablers:

- *Applicable Laws and Regulations*
- *Industry Standards* – relevant industry specifications and standards
- *Agreements* – terms and conditions of the agreements
- *Project Procedures and Standards* – including project plans
- *Organization/Enterprise Policies, Procedures, and Standards* – including guidelines and reporting mechanisms
- *Organization/Enterprise Infrastructure*
- *Project Infrastructure*.

5.2.1.4 Outputs

Outputs of the Project Assessment and Control Process include the following:

- *Project Performance Measures Data* – Data provided to measure project performance
- *Project Status Report* – Information on the health and maturity of the project work effort generated on a periodic basis
- *Project Directives* – Internal project directives based on action required due to deviations from the project plan. New directions are communicated to both project team and customer, when appropriate. If assessments are associated with a decision gate, a decision to proceed, or not to proceed, is taken.
- *Change Requests* – Requests to update any formal baselines that have been established.

5.2.1.5 Process Activities

The Project Assessment and Control Process includes the following activities:

- *Assess the Project*
  - Determine actual and projected cost against budget, actual and projected time against schedule, and deviations in project quality
  - Evaluate the effectiveness and efficiency of the performance of project activities
  - Evaluate the adequacy and the availability of the project infrastructure
  - Evaluate project progress against established criteria and milestones
  - Conduct required reviews, audits, and inspections to determine readiness to proceed to next milestone
  - Monitor critical tasks and new technologies (see Section 5.4, Risk Management)
  - Make recommendations for adjustments to project plans – these are input to the project control process and other decision-making processes
  - Communicate status as designated in agreements, policies, and procedures
  - Analyze assessment results
- *Control the Project*
  - Initiate corrective actions when assessments indicate deviation from approved plans
  - Initiate preventive actions when assessments indicate a trend toward deviation
  - Initiate problem resolution when assessments indicate non-conformance with performance success criteria
  - Establish work items and changes to schedule to reflect the actions taken
  - Negotiate with suppliers for any goods or services acquired from outside the organization
  - Make the decision to proceed, or not to proceed, when assessments support a tollgate or milestone event
- *Close the Project.*

Common approaches and tips:

- One way for project management to remain updated on project status is to conduct regular team meetings. Short standup meetings on a daily or weekly schedule are effective for smaller groups.
- Prevailing wisdom suggests that "what gets measured gets done," but projects should avoid the collection of measures that are not used in decision-making.
- The Project Management Institute provides industry-wide guidelines for project assessment, including Earned Value Management techniques.
- Project teams need to identify critical areas and control them through monitoring, risk management, or configuration management.
- An effective feedback control process is an essential element to enable the improvement of project performance.
- Agile project management techniques schedule frequent assessments and make project control adjustments on tighter feedback cycles than other plan-driven development models.
- Tailoring of organization processes and procedures (see Chapter 8) should not jeopardize any certifications. Processes must be established with effective review, assessment, audit, and upgrade.

5.2.2 Elaboration

Elaborations for this section are under development and will be included in a future revision of the *INCOSE Systems Engineering Handbook*.

5.3 Decision Management Process

5.3.1 Overview

5.3.1.1 Purpose

As stated in ISO/IEC 15288:2008:

> *The purpose of the Decision Management Process is to select the most beneficial course of project action where alternatives exist.*
>
> *This process responds to a request for a decision encountered during the system life cycle, whatever its nature or source, in order to reach specified, desirable or optimized outcomes. Alternative actions are*

analyzed and a course of action selected and directed. Decisions and there rational are recorded to support future decision-making.[9]

Decisions are made throughout the life cycle of every system whenever alternative courses of action exist. Milestones and decision gates mark the most formal decisions. Less formal decisions require less structure, but documenting all decisions, with their rationale, supports future decision-making.

5.3.1.2 Description

As the system progresses from early concept definition throughout sustainment, decisions are needed to direct the focus of all personnel toward the desired result. Every decision involves an analysis of alternative options, the selection of a course of action, and recording of the eventual decision with supporting documentation.

Decisions come from many sources and range from programmatic to highly technical, and different strategies are appropriate to each category of decision. All decisions are taken within the context of an organization. Some decisions are made within the context of other processes, for example, approval of an ECP within the Configuration Management Process (see Section 5.5). Figure 5-5 shows the context diagram for the Decision Management Process.

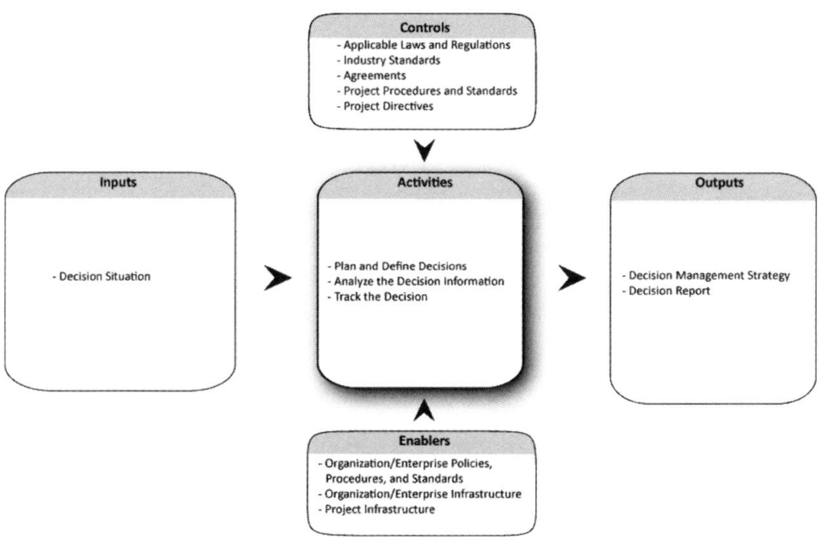

Figure 5-5 Context Diagram for the Decision Management Process

5.3.1.3 Inputs

Inputs to the Decision Management Process include the following:

- *Decision Situation* – Decisions related to decision gates are taken on a pre-arranged schedule; other requests for a decision may arise from any stakeholder and originate from any life cycle process.

This process is governed by the following controls and enablers:

- *Applicable Laws and Regulations*
- *Industry Standards* – relevant industry specifications and standards
- *Agreements* – terms and conditions of the agreements
- *Project Procedures and Standards* – including project plans
- *Project Directives*
- *Organization/Enterprise Policies, Procedures, and Standards* – including guidelines and reporting mechanisms
- *Organization/Enterprise Infrastructure*
- *Project Infrastructure.*

5.3.1.4 Outputs

Outputs of the Decision Management Process include the following:

- *Decision Management Strategy* – Identifies the degrees of rigor and formality regarding decisions
- *Decision Report* – The approved decision is documented and communicated along with rationale, assumptions, constraints, and supporting analysis.

5.3.1.5 Process Activities

The Decision Management Process includes the following activities:

- *Plan and Define Decisions*
  - Identify the need for a decision and the strategy for making the decision, including desired outcomes and measurable success criteria
- *Analyze the Decision Information*
  - Involve all personnel with knowledge and experience relevant to the decision

  - Evaluate the consequences of alternative choices using the selected strategy and optimize the decision
  - Make the decision, based on the relevant data and inputs
- *Track the Decision*
  - Record the decision, with the relevant data and supporting documentation
  - Communicate new directions from the decision.

Common approaches and tips:

- Decisions should be taken in consideration of prior history, and all relevant persons should be involved in the decision-making activities.
- Decision support systems and techniques have been developed to assist decision makers in considering the implications of various courses of action.
- Failure to maintain a history of prior studies and decisions can result in wasted effort when old questions reappear.

5.3.2 Elaboration

Making good decisions requires adequate information, experience, and good judgment. The techniques discussed in the following paragraphs are found in the literature and have proven to be effective aids in making good decisions. In some cases, a technique may use mathematics to produce a result useful in the decision-making process, such as the hydrographical models used to assess the environmental restrictions in the Øresund Strait (see Section 3.6.2). People often make decisions based on intuition and judgment; these techniques are aides to decision-making.[10]

5.3.2.1 Making Difficult Decisions via Decision Analysis

Decision analysis is a method of identifying the best option from a set of alternatives, under uncertainty, using the possible outcomes of each alternative and their probabilities of occurrence to calculate the expected value of the outcome. Decision analysis has been a subject of interest for centuries[11,12,13] and can be applied to a wide-range of problems and problem domains.

David C. Skinner[14] states, "Real world decisions often involve a high degree of ambiguity, conflicting goals due to multiple objectives, complex trade-offs, more than one decision maker, or several sequential decisions. It is these types of situations where decision analysis is most valuable. By carefully decomposing the problem into smaller more manageable problems and by focusing on what is truly important, we can develop clear objectives and defensible courses of action." Skinner also lists ten principles of good decision making:

1. Use a value creation lens for developing and evaluating opportunities
2. Clearly establish objectives and trade-offs
3. Discover and frame the real problem
4. Understand the business situation
5. Develop creative and unique alternatives
6. Identify experts and gather meaningful and reliable information
7. Embrace uncertainty as the catalyst of future performance
8. Avoid "analysis paralysis" situations
9. Use systemic thinking to connect current to future situations
10. Use dialog to foster learning and clarity of action.

Advocates of Lean Manufacturing (see Section 3.4.3) would add one more suggestion, namely, delay commitment until the last responsible moment. Lean software development delays freezing all design decisions as long as possible because it is easier to change a decision that is not made.[15]

Decision trees are a graphical and quantitative method for thinking through a decision. The first step is to create a decision "tree" diagram that represents the situation in question. Starting on the left with the initial decision point and proceeding to the right, the decision diagram must accurately represent each point where a decision is to be made and all possible consequences of that decision. Figure 5-6 illustrates a situation where management needs to decide on whether or not to bid on a contract. The team estimates that their company has a 60% chance of winning. If they propose a unit price of $30, the company will earn $4.25 M. They further estimate that there is a 10% chance that the unit cost will be $28, which would result in an increase in income to $6 M. There is a 40% chance that the unit cost could be as high as $38, and the project will lose about $3 M. The unit costs and probabilities of each chance outcome are based on the best judgment of the team.

The expected value of winning the contract is the sum of the expected value for each branch at the chance node times the probability for each branch, as follows:

Expected Value of Contract Win

10% * \$6.05 M + 50% * \$4.25 M – 40% * \$2.95M, or \$1.55M

Expected Value of Making the Bid

60% * \$1.55 – 40% * 0.25, or \$0.83 M

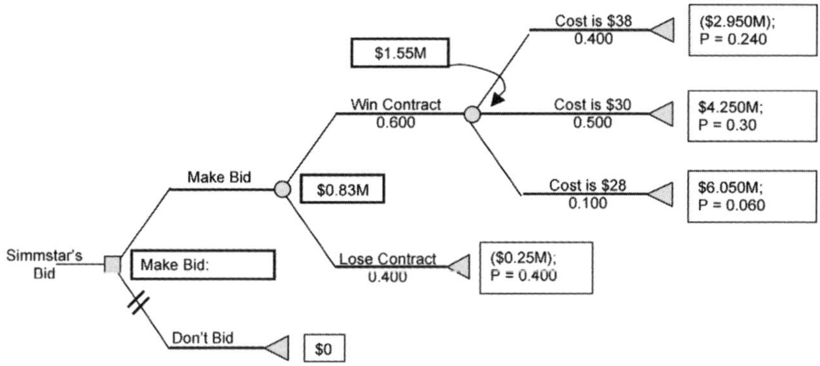

Figure 5-6 Decision Tree for a "Bid – No Bid" Decision[14]

This technique can be extended to include multiple decision points and multiple outcomes as long as every possible outcome has a value and a probability of occurrence associated with it. Additional decision analysis techniques include:

1. *Sensitivity Analysis* — looks at the relationships between the outcomes and their probabilities to find how "sensitive" a decision point is to the relative numerical values.
2. *Value of Information Methods* — whereby expending some effort on data analysis and modeling can improve the optimum expected value.
3. *Multi-attribute Utility Analysis* — develops equivalencies between dissimilar units of measure.

Many tools are available to support the decision management area. The INCOSE web site maintains a current list of suggestions from the Tools Working Group.

5.3.2.2 Trade Studies

Trade studies provide an objective foundation for selecting one of two or more alternative approaches to solve an engineering problem and support decisions in all stages of system development, from conceptualization to deployment. Trade studies may address any of a range of problems from selection of high-level system architecture to selection of a specific COTS processor.

Requirements can be traded against constraints, architecture features can be traded against dictated equipment or interface requirements, and alternative functional or performance choices can be traded to determine an optimal configuration. In the case of the Øresund Bridge (see Section 3.6.2), trade studies helped determine many of the final elements of the bridge configuration (e.g., length of main span).

In developing a design, it is tempting to select a design solution without performing a formal trade study. The selection may seem obvious to the developer as the other possible alternatives appear unattractive, particularly to other team members (e.g., design, manufacturing, quality, and other "-ility" engineering disciplines). However, it will be far easier to justify the selected solution in a proposal or at a formal design review if the team has followed certain procedures in making the selection. Formal trade study procedures will provide discipline in the decision process, and may prevent some ill-advised decisions. It is also important to recognize when a formal trade study is not needed, thus resulting in reduced project costs.

Whenever a decision is made, a trade-off activity is carried out, implicitly, if not explicitly. It is useful to consider trade studies in three levels of formality:

- *Formal* – These trades use a standardized methodology and are formally documented and reviewed with the customer or internally at a design review.
- *Informal* – These trade studies follow the same kind of methodology as formal studies, but are only recorded in the engineer's notebook and are not formally reviewed.
- *Mental* – When a selection of any alternative is made, a mental trade study is implicitly performed with less rigor and formality than documented trades. These types of trade studies are made continuously in our everyday lives. These are appropriate when the consequences of the selection are not too important, when one alternative clearly outweighs all others, or when time does not permit a more extensive trade. However, if the rationale is not documented, it is soon forgotten and unavailable to those who may follow.

One chooses the level of trade study depending on the consequences to the project, the complexity of the issue, and the resources available. The resources to perform trades are allocated based on the overall LCC differences (with provision for risk coverage) in alternative selection for the potential trades. Those with the largest overall LCC deltas are performed first. Since more informal trades can be performed with fewer resources than formal trades, the number and selection of trades and their formality need to be decided with the customer and with the necessary team members who might find some design solutions favorably or unfavorably impacting manufacturability, producibility, reliability, testability, maintainability, etc. Remember, it takes minimal effort to document the rationale for informal and "mental" tradeoff conclusions.

Multiple techniques are available for performing trade studies. These include MAUA, AHP, Decision Trees, and Maximum Expected Utility (MEU). There is no need to standardize on any one technique since one might be better for one trade study and an entirely different one better in another situation. A recent study reported that the following activities can be found in most trade study processes:[16]

1. Frame the decision context, scope, constraints
2. Establish communications with stakeholders
3. Define evaluation criteria (i.e., musts and wants) and weights, where appropriate
4. Define alternatives and select candidates for study
5. Define measures of merit and evaluate selected candidates
6. Analyze the results, including sensitivity analyses, and select best alternative
7. Investigate the consequences of implementation
8. Review results with stakeholders and re-evaluate, if required
9. Use scenario planning to verify assumptions about the future.

With these activities, an objective measure of the suitability of each alternative as a solution to the problem can be obtained. If these activities are performed correctly and objectively, then the alternative with the best score is the best overall solution. Key trade study activities are described in the following paragraphs.

Frame the Decision

The first step in performing a trade study is to clearly articulate the decision that needs to be made. This includes forming a concise statement of the scope and context of the decision being addressed and identifying and documenting any constraints that must be considered in the process.

Determine Screening and Selection Criteria

Systems engineers differentiate between two distinct types of criteria: screening criteria and selection criteria. Screening criteria ("must have") identify those characteristics that are mandatory for any potential solution.

Selection criteria ("want to have") are used to discriminate between viable alternatives based on their respective performance across the entire criteria set.

Screening criteria are used to ensure the viability of alternatives prior to investing additional resources in the trade study. Many times, screening criteria relate directly to key stakeholder requirements and constraints, such as MOEs and MOPs, as discussed in Section 5.7.2.3.[7] Any alternative that does not meet screening criteria is typically not considered further.

In most cases, there should be no difficulty in determining the selection criteria since there are usually key desirable characteristics that are important to the stakeholders. In almost every trade study, cost and risk are certain to be significant factors. Additionally, risk may be decomposed into cost risk, schedule risk, programmatic risk, and performance risk if it appears that these vary separately among the alternatives. Selected performance criteria should be essentially independent and accurately reflect the needs of the system. If the criteria are not independent, they should be kept as a single criterion.

Additionally, decision makers should not overlook life cycle factors that may be significant to the customer. For example, manufacturability may be a key factor. Is the solution maintainable? Is it reliable? Will replacement parts be available in the future? Is the software portable to the platforms that will be available in future years? Is the solution expandable or scalable? Are design elements or software reusable or already available off-the-shelf? Physical parameters, such as size, weight, and power consumption could also be relevant criteria. Where possible, select quantifiable selection criteria that can be used in decision models.

Establish Weighting Values

The weighting values for each criterion reflect their relative importance in the selection process. Values should be assigned in the range of 1 to 10, with 10 applying to the most critical criteria for selection. It is important that all parties interested in the decision reach consensus in the assignment of weights. To achieve objectivity, consensus should be reached before the alternative solutions have been identified.

Establishing weighting values can be a difficult task and can become very subjective. For important trade studies, where weightings are particularly difficult to establish, decision makers should consider using the AHP[17] or Bayesian Team Support.[18]

Identify Viable Alternatives

The next step in performing a trade study is the selection of a number of candidate alternative design solutions. At times, as few as only two alternatives need to be considered. However, in general, the trade study should consider between four and seven reasonable alternatives to ensure that the study does not overlook a viable alternative, while at the same time keeping the cost of the study within reasonable bounds.

Alternatives that cannot meet screening criteria should not be considered further in the trade study. However, if no solution meets the screening criteria, then higher levels of management need to be informed of the problem. In such cases, it might be necessary to include all viable alternatives and assign an appropriate score and weighting value to indicate how closely each alternative comes to meeting the screening criteria.

It is important that alternatives being considered be comparable in completeness (i.e., that one can be substituted in the system directly for the other). Where that is not possible, the selection criteria and weighting values need to take the disparity into account. Alternatives should include those that meet the performance specification but may be more easily produced or more reliable, maintainable, or supportable.

Assign Measures of Merit and Evaluate Alternatives

Measures are assigned to each criterion to characterize how well the various alternatives satisfy the set of selection criteria. This process can be very subjective, and the specific measures and methods used will often vary between practitioners. Systems engineers must use their best engineering judgment in assigning scores and document the rationale and results of the trade study.

A typical trade study scoring approach uses a scale of 1 to 10. A measure of 10 is assigned to the alternative that best meets a particular selection criterion. Other alternatives are scored relative to the best alternative on a scale from 1 to 10. This process is then repeated for each criterion. The subjective component in assigning values arises in determining how to score (i.e., assign values to) various levels of performance. It is essential, however, that there is consistency in how the scores are applied to the various solutions. Two alternatives with the same performance should have the same score for that criterion.

The performance score for each alternative is multiplied by weight (i.e., importance) of each respective criterion to produce a weighted utility value. The weighted total is the sum of the weighted utility values summed over all criteria for a given alternative. The preferred alternative nominally is the one with the best weighted total.

Conduct Sensitivity Analysis

As described above, trade studies compare the appropriateness of different technical solutions by trading the characteristics of each option against each other. Once a best alternative has been identified, stakeholders in the decision will want to know how sensitive the recommended selection is to differently evaluated criteria or to different estimates of the alternatives' characteristics — perhaps a different best alternative would result. Therefore, a good trade study provides a disciplined process that justifies the selected approach, and includes sensitivity analysis to determine if a relatively small variation in scoring affects the outcome.

A sensitivity analysis involves varying each utility and weight and re-computing the weighted total for each alternative to determine what would change if the utility values or weights were different. The significance of the change is best determined through conversations with stakeholders and subject matter experts. All evaluations and decisions are reviewed to address the concerns and opinions of stakeholders. If the decision is based primarily on scoring of an individual factor, that score needs to be given extra care since it essentially determines the selection. The sensitivity analysis should concentrate on the criteria most affecting winner selection. Involvement of the stakeholders in this activity gives them confidence in the eventual choice and imparts useful insights to the whole team.

Determine Adverse Consequences

It is important to consider the adverse consequences that may be associated with the leading alternatives. These adverse consequences may have been reflected in the attributes selected; however, to ensure that they are all considered, a separate step is appropriate. In many cases, where the risk is considerable, this step corresponds to a risk assessment and may be continually tracked as a risk. In any case, the methodology used in performing an adverse consequences analysis is the risk assessment methodology (see Section 5.4). A final evaluation of the consequences of implementing the selected alternative may help identify unintended consequences of an otherwise "best" solution. Hence, the highest score does not always win.

Present the Results

The results of the formal trade study need to be both presented and explained in a report. A summary presentation should include the following:

- Summary description of each alternative solution

- Summary of the evaluation factors used – selection criteria and screening criteria
- Summary of the weighting values used and an explanation of why or how the specific weighting values were selected
- Detailed description of each alternative solution
- Summary description of why or how the specific scores were assigned to each of the alternatives for each of the criteria
- Copy of the analysis spreadsheet (if one was used)
- Graphical display of the overall scores
- Graphical display of the weighted scores for each criterion for each of the alternatives, as shown in Figure 5-7.

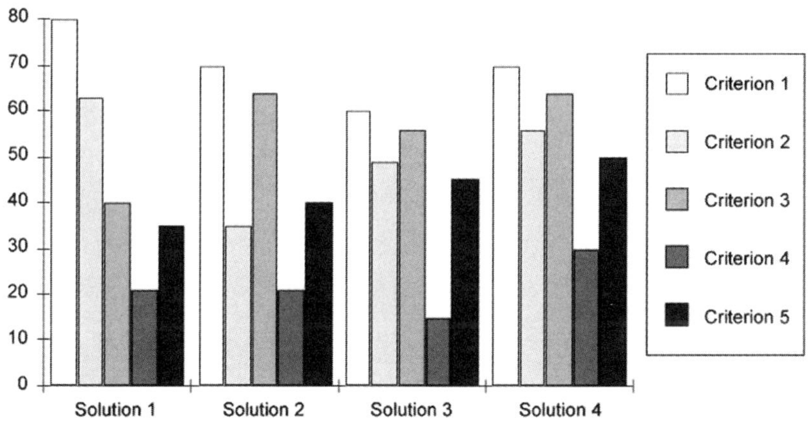

Figure 5-7 Weighted Scores for Each Criterion for Each Alternative

Prepare Formal Trade Study Reports

Trade studies provide visibility into the SE effort and the reasons for selecting one alternative over another. For the most important trades, a report is prepared, and the results are presented at a customer design review. A sample format for a Trade Study Report is shown in Figure 5-8.

1. Scope
2. Tradeoff Study Team Members
   A.
   B. (List Names and Specialties Represented)
   C.
3. Functional and Performance Design Requirements
   A.
   B.
   C.
4. Design Approaches Considered and Significant Characteristics
   A.
   B.
5. Comparison Matrix of the Design Approaches

Feature or Design Requirement	Alternative 1	Alternative 2	Alternative 3	Alternative 4
Requirements 1 (Weight)				
Requirements 2 (Weight)				
Requirements n (Weight)				

6. Design Approach Recommended
   A.
   B.
   C.

Figure 5-8 Sample Trade Study Report Format

The following information is to be included in each of the paragraphs listed in Figure 5-8.

1. *Paragraph 1* – State the scope of the report.

2. *Paragraph 2* – List the names, roles, and any specialties of the team members involved in the trade study.

3. *Paragraph 3* – Identify and list the functional and performance design requirements germane to the trade study. In each subparagraph, state the functional requirement first and then identify the related performance design requirements. Immediately following each requirement (and in the same paragraph), a reference should be made that identifies the source of the requirement. This reference consists of the title, file name or number, date, page number, and paragraph number from which the requirement statement was extracted.

4. *Paragraph 4* – List the possible design approaches and identify the significant characteristics and associated risks of each approach. Only reasonably attainable design approaches should be discussed in detail, considering technical capabilities, time schedules, resource limitations, and requirement constraints.

   Characteristics considered must relate to the attributes of the design approaches bearing most directly on stated requirements. These characteristics should reflect predicted impact on such factors as cost, effectiveness, supportability, personnel selection, training requirements, technical data, schedules, performance, survivability, vulnerability, growth potential, facilities, transportability, and producibility. List the less achievable alternatives with brief statements of why they were not pursued.

5. *Paragraph 5* – Present a matrix comparing the characteristics for each design approach to determine the degree to which the design approaches satisfy the functional and performance design requirements. The objective is to facilitate rapid comparison and evaluation of potential design approaches and to allow preliminary screening out of those design approaches that are inconsistent with the functional and technical design requirements. Where applicable, include cost-effectiveness models and cost analysis data as enclosures.
6. *Paragraph 6* – Recommend the most promising design approach and provide narrative to substantiate the recommendation. Include schematic drawings, outline drawings, interface details, functional diagrams, reliability data, maintainability data, safety data, statistical inference data, and any other documentation or data deemed necessary to support the recommendation. The narrative must cover the requirements that the recommended approach imposes on other areas of the system.

Because there may be a large number of trade study reports prepared during a system development cycle, an index should be prepared that assigns a unique identification number to each trade study report that has been completed.

5.4 Risk Management Process

5.4.1 Overview

5.4.1.1 Purpose

As stated in ISO/IEC 15288:2008:

> *The purpose of the Risk Management Process is to identify, analyze, treat and monitor the risks continuously.*
>
> *The Risk Management Process is a continuous process for systematically addressing risk throughout the life cycle of a system product or service. It can be applied to risks related to the acquisition, development, maintenance or operation of a system.*[19]

5.4.1.2 Description

According to E. H. Conrow, "Traditionally, risk has been defined as the likelihood of an event occurring coupled with a negative consequence of the event occurring. In other words, a risk is a potential problem — something to be avoided if possible, or its likelihood and/or consequences reduced if not."[20] As a corollary, Conrow defines opportunity as "the potential for the realization of wanted, positive consequences of an event."[21]

Risk (and opportunity) management is a disciplined approach to dealing with uncertainty that is present throughout the entire systems life cycle. The objective is to achieve a proper balance between risk and opportunity. This process is used to understand and avoid the potential cost, schedule, and performance (i.e., technical) risks to a system; to take a proactive and structured approach to anticipate negative outcomes and respond to them, if they occur; and to identify potential opportunities that may be hidden in the situation. Organizations manage many forms of risk, and the risk associated with system development is managed in a manner that is consistent with the organization strategy.

Every new system or modification of an existing system is based on pursuit of an opportunity. Risk is always present in the life cycle of systems, and the risk management actions are assessed in terms of the opportunity being pursued. The system may be intended for technical accomplishments near the limits of the state-of-the-art, thus creating technical risk. Risk can also be introduced during architectural design caused by the internal interfaces that exist between the system elements. System development may be rushed to deploy the system as soon as possible to exploit a marketing opportunity or meet an imminent threat, thus leading to schedule risk. All systems are funding-limited so that cost risk is always present. Risk can also develop within a project, since (for example) technical risk can create schedule risk, which in turn can create cost risk.

Ambient risk is often neglected in project management. The ambient risk is defined as the risk caused by and created by the surrounding environment (i.e., ambience) of the project[22]. Project participants have no control over ambient risk factors, but they can learn to observe the external environment and eventually take proactive or reactive actions to minimize the impact of the environment on the project. The typical issues are time dependent processes, rigid sequence of activities, one dominant path for success, and little slack.

Projects are also subject to uncertainty. An uncertain event may be harmful if it occurs (i.e., threats), another may assist in achieving objectives (i.e., opportunities). Dealing with both types of uncertainty under the single heading of "risk management" minimizes process and overhead and expands organizational and personal commitment towards finding and capturing opportunities. Since project managers traditionally think of risks as threats alone, it may be a change to begin recognizing opportunities as risks. If opportunities are treated along with threats, risk management language needs to be balanced with terms for opportunities, such as "exploit," "share," "protect," and "enhance." Project managers may need encouragement to be open to opportunities and to manage both threats and opportunities proactively.

Typical strategies for coping with risk include transference, avoidance, acceptance, or taking action to reduce the anticipated negative effects of the situation. Most Risk Management Processes include a prioritization scheme whereby risks with the greatest negative effect and the highest probability of occurrence are treated before those deemed to have lower negative consequences and lower probability of occurrence. The objective of risk management is to balance the allocation of resources such that the minimum amount of resources achieves the greatest risk mitigation (or opportunity realization) benefits. Figure 5-9 shows the context diagram for the Risk Management Process.

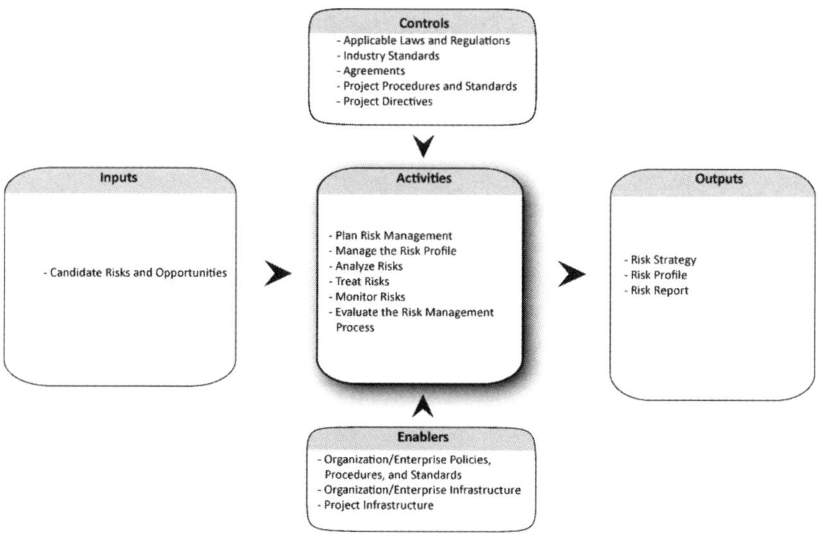

Figure 5-9 Context Diagram for the Risk Management Process

5.4.1.3 Inputs

The inputs to the Risk Management Process include the following:

- *Candidate Risks and Opportunities* – Risks and opportunities may arise from any stakeholder and originate from any life-cycle process. In many cases, risk situations are identified during the Project Assessment and Control Process.

This process is governed by the following controls and enablers:

- *Applicable Laws and Regulations*
- *Industry Standards* – relevant industry specifications and standards
- *Agreements* – terms and conditions of the agreements
- *Project Procedures and Standards* – including project plans
- *Project Directives*
- *Organization/Enterprise Policies, Procedures, and Standards* – including guidelines and reporting mechanisms
- *Organization/Enterprise Infrastructure*
- *Project Infrastructure.*

5.4.1.4 Outputs

Outputs of the Risk Management Process include the following:

- *Risk Management Strategy*
- *Risk Profile* – Sometimes referred to as a risk matrix; it contains the findings of the Risk Management Process
- *Risk Report* – The risks are documented and communicated along with rationale, assumptions, treatment plans, and current status. For selected risks, an action plan is produced to direct the project team to properly respond to the risks. If appropriate, change requests are generated to mitigate technical risk.

5.4.1.5 Process Activities

The Risk Management Process includes the following activities:

- *Plan Risk Management*
  - Define and document the risk strategy
- *Manage the Risk Profile*
  - Define and document risk thresholds and acceptable and unacceptable risk conditions
  - Periodically communicate the risks (and opportunities) with the appropriate stakeholders

- *Analyze Risks*
  - Identify and define risk situations
  - Analyze risks for likelihood and consequence to determine the magnitude of the risk and its priority for treatment
  - Define a treatment scheme and resources for each risk, including identification of a person who will be responsible for continuous assessment of the status of the situation
- *Treat Risks*
  - Using the criteria for acceptable and unacceptable risk, generate a plan of action when the risk threshold exceeds acceptable levels
- *Monitor Risks*
  - Maintain a record of risk items and how they were treated
  - Maintain transparent risk management communications
- *Evaluate the Risk Management Process*
  - Define, analyze, and document measures indicating the status of the risk and the effectiveness of the treatment alternatives.

Common approaches and tips:

- In the Project Planning Process (see Section 5.1), a Risk Management Plan (RMP) is tailored to satisfy the individual project procedures for risk management.
- A Risk Management Process establishes documentation, maintained as the Risk Profile, that includes a description, priority, mitigation, responsible person, and status of each risk item.
- One rule of thumb for identifying risks is to pose each risk candidate in an "if <situation>, then <consequence>" format. This form helps to determine the validity of a risk and assess its magnitude or importance. If the statement does not make sense or cannot be put in this format, then the candidate is probably not a true risk. For example, a statement that describes a situation but not a consequence implies that the potential event will not affect the project. Similarly, a statement of potential consequence without a clear situation description is worthy of more attention.
- Document everything so if unforeseen issues and challenges arise during execution the project can recreate the environment within

which the planning decisions were made and know where to update the information to correct the problem.

- Negative feedback toward personnel who identify a potential problem will discourage the full cooperation of engaged stakeholders and could result in failure to address serious risk-laden situations. Conduct a transparent Risk Management Process to encourage suppliers and other stakeholders to assist in risk mitigation efforts. Some situations can be difficult to categorize in terms of probability and consequences; involve all relevant stakeholders in this evaluation to capture the maximum variety in viewpoints.
- Many analyses completed throughout the Technical Processes, such as FMECA, may identify candidate risk elements.
- The measures for risk management vary by organization and by project. As with any measure, use measurement analysis or statistics that help manage the risk.
- The Project Management Institute is a good source for more information on Risk Management.
- The Institute of Risk Management has generated *The Risk Management Standard*[23] to assist organizations in risk management.
- Forsberg, et al., *Visualizing Project Management*, contains additional reading on opportunity management.[24]
- ISO/IEC 16085:2006, *Systems and software engineering – Life cycle processes – Risk management*, also provides more detail on the Risk Management Process.[25]

5.4.2 Elaboration

5.4.2.1 Risk Management Concepts

Most projects are executed in an environment of uncertainty. Risks (also called threats) are events that if they occur can influence the ability of the project team to achieve project objectives and jeopardize the successful completion of the project.[26] Well-established techniques exist for managing threats, but there is some debate over whether the same techniques are applicable to recognizing opportunities. In an optimal situation, opportunities are maximized at the same time as threats are minimized, resulting in the best chance to meet project objectives.[27] The Øresund Bridge case (see Section 3.6.2) illustrates this in that the man-made Peberholm Island was created from the materials dredged from the Strait to meet environmental requirements and is now a sanctuary for a rare species of tern.

The measurement of risk has two components (see Figure 5-10):

- The likelihood that an event will occur
- The undesirable consequence of the event if it does occur.

The likelihood that an undesirable event will occur is often expressed as a probability. The consequence of the event is expressed in terms that depend on the nature of the event (e.g., lost investment, inadequate performance, etc.). The combination of low likelihood and low undesirable consequences gives low risk, while high risk is produced by high likelihood and highly undesirable consequences.

By changing the adjective from undesirable to desirable the noun changes from risk to opportunity, but the diagram remains the same. As suggested by the shading, most projects experience a comparatively small number of high risk or high opportunity events.

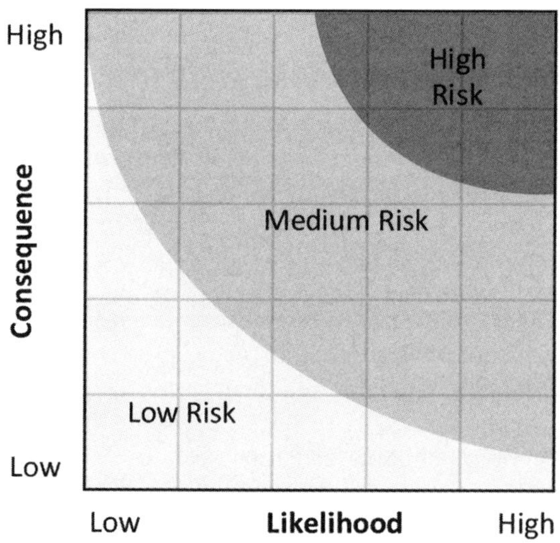

Figure 5-10 Level of risk depends upon both likelihood and consequences

There is no alternative to the presence of risk in system development. The only way to remove risk is to set technical goals very low, to stretch the schedule, and to supply unlimited funds. None of these events happen in the real world, and no realistic project can be planned without risk. The challenge is to define the system and the project that best meet overall requirements, allow for risk, and achieve the highest chances of project success. Figure 5-11 illustrates the major interactions between the four risk categories: technical, cost, schedule and programmatic. The arrow labels indicate typical risk relationships, though others certainly are possible.

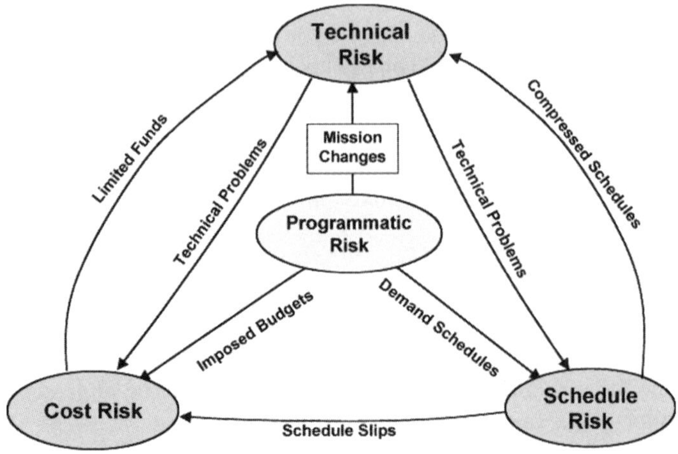

Figure 5-11 Typical Relationship among the Risk Categories

- *Technical risk* – the possibility that a technical requirement of the system may not be achieved in the system life cycle. Technical risk exists if the system may fail to achieve performance requirements; to meet operability, producibility, testability, or integration requirements; or to meet environmental protection requirements. A potential failure to meet any requirement that can be expressed in technical terms is a source of technical risk.

- *Cost risk* – the possibility that available budget will be exceeded. Cost risk exists if the project must devote more resources than planned to achieve technical requirements, if the project must add resources to support slipped schedules due to any reason, if changes must be made to the number of items to be produced, or if changes occur in the organization or national economy. Cost risk can be predicted at the total project level or for a system element. The collective effects of element-level cost risks can produce cost risk for the total project.

- *Schedule risk* – the possibility that the project will fail to meet scheduled milestones. Schedule risk exists if there is inadequate allowance for acquisition delays. Schedule risk exists if difficulty is experienced in achieving scheduled technical accomplishments, such as the development of software. Schedule risk can be incurred at the total project level for milestones such as deployment of the first system element. The cascading effects of element-level schedule risks can produce schedule risk for the total project.

- *Programmatic risk* – produced by events that are beyond the control of the project manager. These events often are produced by decisions made by personnel at higher levels of authority, such as reductions in project priority, delays in receiving authorization to proceed with a project, reduced or delayed funding, changes in organization or national objectives, etc. Programmatic risk can be a source of risk in any of the other three risk categories.

5.4.2.2 Risk Management Approach

Once a risk management strategy and risk profile have been established, the three key Risk Management Process activities are: Analyze Risks, Treat Risks, and Monitor Risks.

Analyze Risks

Analyzing risks involves identifying risks and evaluating their relative likelihood and consequence. The basis for this evaluation may be qualitative or quantitative; regardless, the objective is to set priorities and focus attention on areas of risk with the greatest consequences to the success of the project. All stakeholders and project personnel should feel welcome to contribute to identifying and analyzing risks.

If a project is unprecedented, brainstorming using Strength-Weakness-Opportunity-Threat (SWOT) or Delphi techniques may be appropriate. However, most projects represent a new combination of existing systems or system elements or represent the insertion of incremental advances in technology. This means that key insights can be gained concerning a current project's risk by examining the successes, failures, problems, and solutions of similar prior projects. The experience and knowledge gained, or lessons learned, can be applied to identify potential risk in a new project and to develop risk-specific management strategies.

The first step is to determine the information needs. This could vary from assessing the risk in development of a custom computer chip to identifying the risks associated with a major system development. Next, systems engineers define the basic characteristics of the new system as a basis for identifying past projects that are similar in technology, function, design, etc. Based on the availability of data, analogous systems or subsystems are selected and data gathered. Often the data collection process and initial assessment lead to a further definition of the system for the purposes of comparison. Comparisons to prior systems may not be exact or the data may need to be adjusted to be used as a basis for estimating the future. The desired output is insight into cost, schedule, and technical risks of a project based on observations of similar past projects.

Uncertainty is characterized by a distribution of outcomes based on likelihood of occurrence and severity of consequences. As noted above, risk involves both the likelihood and consequences of the possible outcomes. In its most general form, risk analysis should capture the spectrum of outcomes relative to the desired project technical performance, cost, and schedule requirements. Risk generally needs to be analyzed subjectively because adequate statistical data are rarely available. Expert interviews and models are common techniques for conducting risk analyses.

Expert Interviews

Efficient acquisition of expert judgments is extremely important to the overall accuracy of the risk management effort. The expert interview technique consists of identifying the appropriate experts, questioning them about the risks in their area of expertise, and quantifying these subjective judgments. One result is the formulation of a range of uncertainty or a probability density (with respect to cost, schedule, or performance) for use in any of several risk analysis tools.

Since expert interviews result in a collection of subjective judgments, the only real "error" can be in the methodology for collecting the data. If it can be shown that the techniques for collecting the data are not adequate, then the entire risk assessment can become questionable. For this reason, the methodology used to collect the data must be thoroughly documented and defensible. Experience and skill are required to encourage the expert to divulge information in the right format. Typical problems encountered include identification of the wrong expert, obtaining poor quality information, unwillingness of the expert to share information, changing opinions, getting biased viewpoints, obtaining only one perspective, and conflicting judgments. When conducted properly, expert interviews provide reliable qualitative information. However, the transformation of that qualitative information into quantitative distributions or other measures depends on the skill of the analyst.

Models

Risk is often expressed only in qualitative terms or by a single value. Therefore, it is important to quantify risk in some methodical way to ensure a good allocation of resources for risk reduction. Ideally, risk would be characterized using cumulative probability curves with the probability of failure and the consequences expressed quantitatively in measurable terms. However, given the inherent lack of data and limited analysis, this is usually impractical. It is important to properly quantify risk because an invalid assessment could lead to an improper conclusion with misapplication of resources.

A somewhat subjective, relative rating of risk is developed using an *Expected Value Model*, where risk is expressed as:

Risk = Probability of failure (Pf) * Consequence of failure (Cf)

For illustration purposes, consider a proposal to develop a new light-weight, compact power supply with an operating life of 8,000 hours. The consequences of failing to meet at least 6,000 hours are assessed to be critical, so the *Cf* is assigned a value of 0.8. Given the present state of technology, cost, and schedule, the probability of failing to achieve an operating life of 6,000 hours is judged to be relatively low and is estimated as 30%, so the *Pf* is assigned a value of 0.3. Applying the above equation to this example yields:

Risk = 0.3*0.8 = 0.24

This would suggest a relatively low risk situation. Intuitively, the described scenario represents a low/moderate risk (subjective judgment); therefore this approach appears to yield a valid relative ranking of risk.

Treat Risks

Risk treatment approaches (also referred to as risk handling approaches) need to be established for the moderate and high-risk items identified in the risk analysis effort. These activities are formalized in the RMP. There are four basic approaches to treat risk:

- Avoid the risk through change of requirements or redesign
- Accept the risk and do no more
- Control the risk by expending budget and other resources to reduce likelihood and/or consequence
- Transfer the risk by agreement with another party that it is in their scope to mitigate.

The following steps can be taken to avoid or control unnecessary risks:

- *Requirements scrubbing* – Requirements that significantly complicate the system can be scrutinized to ensure that they deliver value equivalent to their investment. Find alternative solutions that deliver the same or comparable capability.
- *Selection of most promising options* – In most situations, several options are available. A trade study can include project risk as a criterion when selecting the most promising alternative.
- *Staffing and team building* – Projects accomplish work through people. Attention to training, teamwork, and employee morale can help avoid risks introduced by human errors.

For high-risk technical tasks, risk avoidance is insufficient and can be supplemented by the following approaches:

- Early procurement
- Initiation of parallel developments
- Implementation of extensive analysis and testing
- Contingency planning.

The high-risk technical tasks generally imply high schedule and cost risks. Cost and schedule are impacted adversely if technical difficulties arise and the tasks are not achieved as planned. Schedule risk is controlled by early procurement of long-lead items and provisions for parallel-path developments. However, these activities also result in increased early costs. Testing and analysis can provide useful data in support of key decision points. Finally, contingency planning involves weighing alternative risk mitigation options.

For each risk that is determined credible after analysis, a Risk Treatment Plan (also referred to as a Risk Mitigation Action Plan) should be created that identifies the risk treatment strategy, the trigger points for action, and any other information to ensuring the treatment is effectively executed. The Risk Treatment Plan can be part of the risk record on the risk profile. For risks that have significant consequences, a contingency plan should be created in case the risk treatment is not successful. It should include the triggers for enacting a contingency plan.

In China, the authorities built the short Maglev train line in Shanghai (see Section 3.6.3) as a proof-of-concept. In spite of the high investment, this represented lower risk to the project than attempting a longer line with an unproven technology. The results collected from this project are inspiring others to consider Maglev alternatives for greater distances.

Monitor Risks

Project management uses measures to simplify and illuminate the Risk Management Process. Each risk category has certain indicators that may be used to monitor project status for signs of risk. Tracking the progress of key system technical parameters can be used as an indicator of technical risk.

The typical format in tracking technical performance is a graph of a planned value of a key parameter plotted against calendar time. A second contour showing actual value achieved is included in the same graph for comparative purposes. Cost and schedule risk are monitored using the products of the Cost/Schedule Control System or some equivalent technique. Normally, cost and schedule variances are used along with a comparison of tasks planned to tasks accomplished. A number of additional references exist on the topic of risk management.[28,29,30,31,32]

5.4.2.3 Opportunity Management Concepts[33]

Systems engineering and project management are all about pursuing an opportunity to solve a problem or fulfill a need. Opportunities enable creativity in resolving concepts, architectures, designs, strategic and tactical approaches, as well as the many administrative issues within the project. It is the selection and pursuit of these strategic and tactical opportunities that determine just how successful the project and system will be. Of course, opportunities usually carry risks, and each opportunity will have its own set of risks that must be intelligently judged and properly managed to achieve the full value.

Opportunities represent the potential for improving the value of the project results. The project champions (e.g., the creators, designers, integrators, and implementers) apply their "best-in-class" practices in pursuit of opportunities. After all, the fun of working on projects is doing something new and innovative. It is these opportunities that create the project's value. *Risks* are defined as chances of injury, damage, or loss. Risks are the chances of not achieving the results as planned. Each of the strategic and tactical opportunities pursued have associated risks that undermine and detract from the opportunity's value. These are the risks that must be managed to enhance the opportunity value and the overall value of the project (see Figure 5-12). Opportunity and risk management are, therefore, essential to—and performed concurrently with—the planning process but require the application of separate and unique techniques that justify this distinct technical management element.

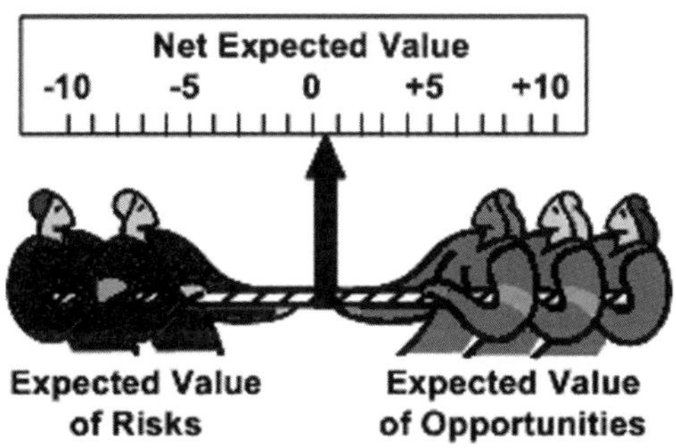

Figure 5-12 Intelligent Management of Risks and Opportunities

There are two levels of opportunities and risks. Because a project is the pursuit of an opportunity, the *macro* level is the project opportunity itself. The approach to achieving the macro opportunity and the mitigation of associated project-level risks are structured into the strategy and tactics of the project cycle, the selected decision gates, the teaming arrangements, key personnel selected, and so on. The *element* level encompasses the tactical opportunities and risks within the project that become apparent at lower levels of decomposition and as project life cycle stages are planned and executed. This can include emerging, unproven technology; incremental and evolutionary methods that promise high returns; and the temptation to circumvent proven practices to deliver products better, faster, and cheaper.

Overall project value can be expressed as benefit divided by cost. Opportunities and their risks should be managed jointly to enhance project value. This is based on the relative merits of exploiting each opportunity and mitigating each risk. In the context of the opportunity and the resultant value, we carry a spare tire to mitigate the risk of a flat tire by reducing the probability and impact of having a delayed trip. The high value we place on getting where we want to go far exceeds the small expense of a spare. When deciding to pursue the opportunity of a long automobile trip, we may take extra risk management precautions, such as preventive maintenance and spares for hard-to-find parts.

The assessment of opportunity and risk balance is situational. For example, few today have a car with more than one spare tire (multiple spares were common practice in the early 1900s). However, a few years ago an individual decided to spend a full month driving across the Australian Outback in late spring. He was looking for solitude in the wilderness (his opportunity). On advice from experienced friends, he took four spare tires and wheels. They also advised him that the risk of mechanical breakdown was very high on a 30-day trip, and the consequence would almost certainly be fatal. The risk of two vehicles breaking down at the same time was acceptably low. So, he adjusted the opportunity for absolute solitude by joining two other adventurers. They set out in three cars. Everyone survived in good health, but only two cars returned, and two of his "spare" tires were shredded by the rough terrain. The "balanced" mitigation approach proved effective.

5.5 Configuration Management Process

5.5.1 Overview

5.5.1.1 Purpose

As stated in ISO/IEC 15288:2008:

> *The purpose of the Configuration Management Process is to establish and maintain the integrity of all identified outputs of a project or process and make them available to concerned parties.*[34]

This is accomplished by ensuring the effective management of the evolving configuration of a system, both hardware and software, during its life cycle. Fundamental to this objective is the establishment, control, and maintenance of software and hardware baselines. Baselines are business, budget, functional, performance, and physical reference points for maintaining development and control. These baselines, or reference points, are established by review and acceptance of requirements, design, and product specification documents. The creation of a baseline may coincide with a project milestone or decision gate. As the system matures and moves through the life-cycle stages, the software or hardware baseline is maintained under configuration control.

5.5.1.2 Description

Evolving system requirements are a reality that must be addressed over the life of a system development effort, and throughout the Utilization and Support Stages of the system. Configuration Management ensures that product functional, performance, and physical characteristics are properly identified, documented, validated, and verified to establish product integrity; that changes to these product characteristics are properly identified, reviewed, approved, documented, and implemented; and that the products produced against a given set of documentation are known. Figure 5-13 presents the context diagram for the Configuration Management Process.

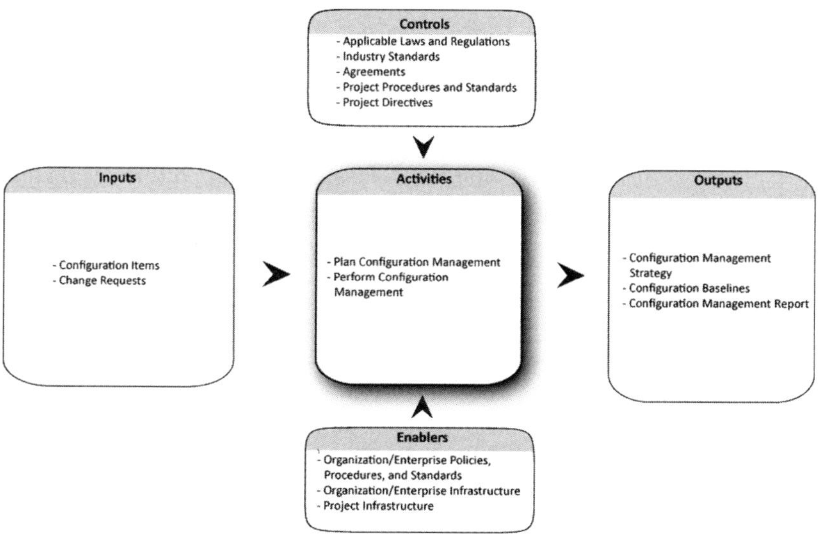

Figure 5-13 Context Diagram for the Configuration Management Process

5.5.1.3 Inputs

Inputs to the Configuration Management Process include the following:

- *Configuration Items (CIs)* – Items for configuration control can originate from any life cycle process
- *Change Requests* – Can originate from any life cycle process, in many cases, the need for change requests is identified during the project assessment process.

This process is governed by the following controls and enablers:

- *Applicable Laws and Regulations*
- *Industry Standards* – relevant industry specifications and standards
- *Agreements* – terms and conditions of the agreements
- *Project Procedures and Standards* – including project plans
- *Project Directives*
- *Organization/Enterprise Policies, Procedures, and Standards* – including guidelines and reporting mechanisms
- *Organization/Enterprise Infrastructure*
- *Project Infrastructure.*

5.5.1.4 Outputs

Outputs of the Configuration Management Process include the following:

- *Configuration Management Strategy*
- *Configuration Baselines* – Places items under formal change control. The required configuration baseline documentation is developed and approved in a timely manner to support required SE technical reviews, the system's acquisition and support strategies, and production.
- *Configuration Management Report* – Documents the impact to any process, organization, decision (including any required change notification), products, and service affected by a given change request.

5.5.1.5 Process Activities

The Configuration Management Process includes the following activities:

- *Plan Configuration Management*
  - Implement a configuration control cycle that incorporates evaluation, approval, validation, and verification of engineering change requests (ECRs)

- *Perform Configuration Management*
  - Configuration Identification – Identify system elements to be maintained under configuration control
  - Configuration Control – Establish the configuration baselines and control baseline changes throughout the system life cycle
  - Configuration Status Accounting – Develop and maintain configuration control documentation and communicate the status of controlled items to the project team
  - Configuration Audits – Perform audits associated with milestones and decision gates to validate the baselines.

Common approaches and tips:

- In the Project Planning Process (see Section 5.1), a Configuration Management Plan (CMP) is tailored to satisfy the individual project procedures for configuration management.
- The primary output of the Configuration Management Process is the maintenance of the configuration baseline for the system and system elements wherein items are placed under formal control as part of the decision-making process.
- Establish a CCB with representation from all stakeholders and engineering disciplines participating on the project.
- Begin the Configuration Management Process in the infancy stages of the system and continue through until disposal of the system.
- Configuration management documentation is maintained throughout the life of the system.

5.5.2 Elaboration

5.5.2.1 Configuration Management Concepts

Change is inevitable, as indicated in Figure 5-14. The purpose of Configuration Management is to establish and maintain control of requirements, documentation, and artifacts produced throughout the system's life cycle and to manage the impact of change on a project. Systems engineers ensure that the change is necessary, and that the most cost-effective recommendation has been proposed.

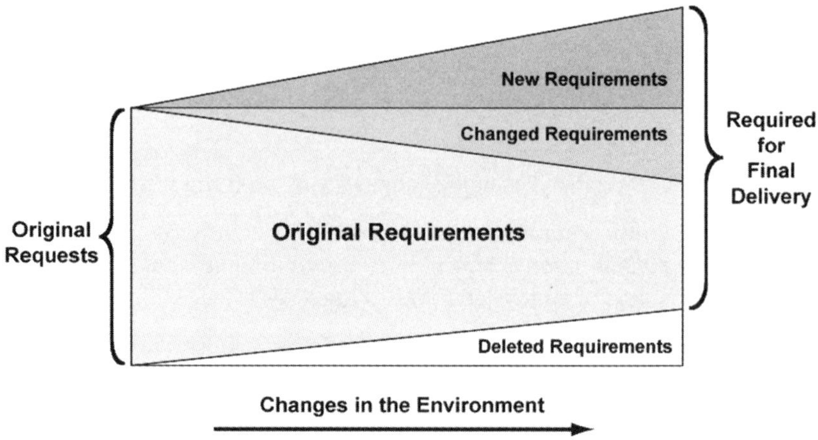

Figure 5-14 Requirements changes are inevitable

Initial planning efforts for configuration management are performed at the onset of the project and defined in the CMP, which establishes the scope of items that are covered by the plan; identifies the resources and personnel skill level required; defines the tasks to be performed; describes organizational roles and responsibilities; and identifies configuration management tools and processes, as well as methodologies, standards, and procedures that will be used on the project. Configuration control maintains integrity by facilitating approved changes and preventing the incorporation of unapproved changes into the items under configuration control. Such activities as check-in and check-out of source code, versions of system elements, and deviations of manufactured items are part of configuration management. Independent configuration audits assess the evolution of a product to ensure compliance to specifications, policies, and contractual agreements. Formal audits may be performed in support of decision gate review.

A request to change the current configuration of a system is typically made using an ECP. An ECP may originate in a number of ways. The customer may request an ECP to address a change in requirements or a change in scope; an unexpected breakthrough in technology may result in the supplier of a system element proposing an ECP; or a supplier may identify a need for changes in the system under development. Circumstances like these that will potentially change the scope or the requirements are appropriate reasons to propose an ECP and to conduct an analysis to understand the effect of the change on existing plans, costs, and schedules. The ECP must be approved before the change is put into effect. It is never appropriate to propose an ECP to correct cost or schedule variances absent of change in scope. A minor change that falls within the current project scope usually does not require an ECP but should be approved and result in the generation of an engineering notice. It is also important to ask, "What is the impact of not making the change?" especially as the system matures, since changes made later in the life cycle have an increasing risk of hidden impacts, which can adversely affect system cost, schedule, and technical performance.

The most desirable outcomes of an ECP cycle are:

1. System functionality is altered to meet a changing requirement
2. New technology or a new product extends the capabilities of the system beyond those initially required in ways that the customer desires
3. The costs of development, or of utilization, or of support are reduced
4. The reliability and availability of the system are improved.

Outcomes 3 and 4 reduce LCCs and potentially save more money than is invested to fund the proposed change.

ECPs and ENs help ensure that a system evolves in ways that allow it to continue to satisfy its operational requirements and its objectives and that any modification is known to all relevant personnel. The camera system illustrated in Figure 2-3 is an example of a product family that depends on accurate identification of system elements and characteristics to support the mix and match consumer market.

5.5.2.2 Configuration Management Approach

Configuration Management establishes and maintains control over requirements, specifications, configuration definition documentation, and design changes. Configuration Identification, Configuration Control, Configuration Status Accounting, and Configuration Audits of the functional and physical configuration (i.e., validation and distribution) are the primary configuration management.

There will always be a need to make changes; however, SE must ensure (1) that the change is necessary and (2) that the most cost-effective solution has been proposed. Configuration Management must, therefore, apply technical and administrative direction, surveillance, and services to do the following:

- Identify and document the functional and physical characteristics of individual CIs such that they are unique and accessible in some form
- Assign a unique identifier to each version of each CI
- Establish controls to allow changes in those characteristics

- Concur in product release and ensure consistent products via the creation of baseline products
- Record, track, and report change processing and implementation status and collect measures pertaining to change requests or problems with the product baseline
- Maintain comprehensive traceability of all transactions.

Configuration Identification

Configuration identification uniquely identifies the elements within a baseline configuration. This unique identification promotes the ability to create and maintain master inventory lists of baselines. As part of the SE effort, the system is decomposed into CIs, which serve as the critical elements subjected to rigorous formal control. The compilation of all the CIs is called the CI list. This list may reflect items that are developed, vendor produced, or provided by the customer for integration into the final system. These items may be deliverable items under the contract or used to produce the deliverable items.

Configuration Control

Configuration control manages the collection of the items to be baselined and maintains the integrity of the CIs identified by facilitating approved changes (e.g., via ECRs) and preventing the incorporation of unapproved changes into the baseline. Change control should be in effect beginning at project initiation.

Change Classification

Effective configuration control requires that the extent of analysis and approval action for a proposed engineering change be in concert with the nature of the change. The problem statement includes a description of the proposed change, the reason for the proposed change, the impacts of the change on cost and schedule, and all affected documentation. Change classification is a primary basis of configuration control. All changes to baselined documents are classified as outside of the scope of the requirements or within the scope of the requirements. A change outside the scope of project requirements is a change to a project baseline document that affects the form, fit, specification, function, reliability, or safety. The coordinating review board determines if this proposed change requires a change notice for review and approval.

Changes are sometimes categorized into two main classes: Class I and Class II. A Class I change is a major or significant change that affects cost, schedule, or technical performance. Normally, Class I changes require customer approval prior to being implemented. A Class II change is a minor change that often affects documentation errors or internal design details. Generally, Class II changes do not require customer approval.

Configuration Control Board

An overall CCB is implemented at project initiation to provide a central point to coordinate, review, evaluate, and approve all proposed changes to baselined documentation and proposed changes to baselined configurations, including hardware, software, and firmware. The review board is composed of members from the various disciplines, including SE, software and hardware engineering, project management, product assurance, and configuration management. The chairperson is delegated the necessary authority to act on behalf of the project manager in all matters falling within the review board responsibilities. The Configuration Management organization is delegated responsibility for maintaining status of all proposed changes. Satellite or subordinate boards may be established for reviewing software or hardware proposed changes below the CI level. If those changes require a higher approval review, they are forwarded to the overall review board for adjudication.

Changes that fall within the review board jurisdiction should be evaluated for technical necessity, compliance with project requirements, compatibility with associated documents, and project impact. As changes are written while the hardware and/or software products are in various stages of manufacture or verification, the review board should require specific instructions for identifying the affectivity or impact of the proposed software or hardware change and disposition of the in-process or completed hardware and/or software product. The types of impacts the review board should assess typically include that the following:

- All parts, materials, and processes are specifically approved for use on the project
- The design depicted can be fabricated using the methods indicated
- Project quality and reliability assurance requirements are met
- The design is consistent with interfacing designs.

Methods and Techniques

Change control forms provide a standard method of reporting problems and enhancements that lead to changes in formal baselines and internally controlled items. The following forms provide an organized approach to changing hardware, software, or documentation:

- *Problem/Change Reports* – can be used for documenting problems and recommending enhancements to hardware/software or its complementary documentation. These forms can be used to identify problems during design, development, integration, verification, and validation.

- *Specification Change Notice (SCN)* – used to propose, transmit, and record changes to baselined specifications.
- *Engineering Change Proposals* – used to propose Class I changes to the customer. These proposals describe the advantages of the proposed change and available alternatives, and identify funding needed to proceed.
- *Engineering Change Requests (ECR)* – used to propose Class II changes.
- *Request for Deviation/Waiver* – used to request and document temporary deviations from configuration identification requirements when permanent changes to provide conformity to an established baseline are not acceptable.

Configuration Status Accounting

Status accounting is performed by the Configuration Management organization to record and report information to management. Configuration management maintains a status of approved documentation that identifies and defines the functional and physical characteristics, status of proposed changes, and status of approved changes. This subprocess synthesizes the output of the identification and control subprocesses. All changes authorized by the configuration review boards (both overall and subordinate) culminate in a comprehensive traceability of all transactions. Such activities as check-in and check-out of source code, builds of CIs, deviations of manufactured items, waiver status are part of the status tracking. By statusing and tracking project changes, a gradual change from the *build-to* to the *as-built* configuration is captured. Suggested measures for consideration include the following:

- Number of changes processed, adopted, rejected, and open
- Status of open change requests
- Classification of change requests summary
- Number of deviations or waivers by CI
- Number of problem reports open, closed, and in-process
- Complexity of problem reports and root cause
- Labor associated with problem resolution and verification stage when problem was identified
- Processing times and effort for deviations, waivers, ECPs, SCNs, ECRs, and Problem Reports
- Activities causing a significant number of Change Requests; and rate of baseline changes.

Configuration Audits

Configuration audits are performed independently by configuration management and product assurance to evaluate the evolution of a product and ensure compliance to specifications, policies, and contractual agreements. Formal audits, or Functional and Physical Configuration Audits, are performed at the completion of a product development cycle.

The Functional Configuration Audit is intended to validate that the development of a CI has been completed and has achieved the performance and functional characteristics specified in the System Specification (functional baseline). The Physical Configuration Audit is a technical review of the CI to verify the as-built maps to the technical documentation. Finally, configuration management performs periodic in-process audits to ensure that the Configuration Management Process is followed.

5.6 Information Management Process

5.6.1 Overview

5.6.1.1 Purpose

As stated in ISO/IEC 15288:2008:

> *The purpose of the Information Management Process is to provide relevant, timely, complete, valid and, if required, confidential information to designated parties during and, as appropriate, after the system life cycle.*
>
> *This process generates, collects, transforms, retains, retrieves, disseminates and disposes of information. It manages designated information, including technical, project, organizational, agreement and user information.*[35]

Information Management ensures that information is properly stored, maintained, secured, and accessible to those who need it, thereby establishing/maintaining integrity of relevant system life-cycle artifacts.

5.6.1.2 Description

Information exists in many forms, and different types of information have different value within an organization. Information assets, whether tangible or intangible, have become so pervasive in contemporary organizations that they are indispensable. The impact of threats to secure access, confidentiality, integrity, and availability of information can cripple the ability to get work done. As information systems become increasingly interconnected, the opportunities for compromise increase.[36] The following are important terms in Information Management:

- Information is what an organization has compiled or its employees know. It can be stored and communicated, and it might include customer information, proprietary information, and/or protected (e.g., by copyright, trademark, or patent) and unprotected (e.g., business intelligence) Intellectual Property.
- Information assets are intangible information and any tangible form of its representation, including drawings, memos, e-mail, computer files, and databases.
- Information security generally refers to the confidentiality, integrity, and availability of the information assets.
- Information security management includes the controls used to achieve information security and is accomplished by implementing a suitable set of controls, which could be policies, practices, procedures, organizational structures, and software.
- Information Security Management System is the life-cycle approach to implementing, maintaining, and improving the interrelated set of policies, controls, and procedures that ensure the security of an organization's information assets in a manner appropriate for its strategic objectives.

Information management provides the basis for the management of and access to information throughout the system life cycle, including after disposal if required. Designated information may include organization/enterprise, project, agreement, technical, and user information. The mechanisms for maintaining historical knowledge in the prior processes – decision-making, risk, and configuration management – are under the responsibility of information management. Figure 5-15 shows the context diagram for the Information Management Process.

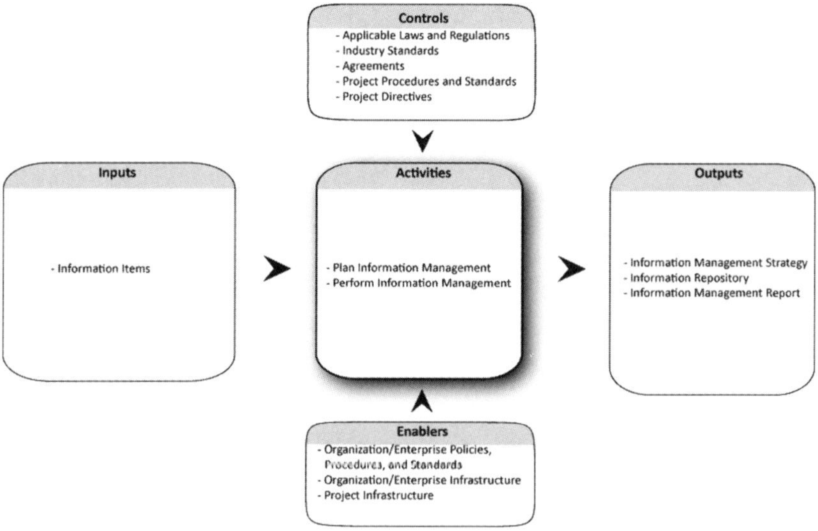

Figure 5-15 Context Diagram for the Information Management Process

5.6.1.3 Inputs

The inputs to the Information Management Process include the following:

- *Information Items* –Information items can originate from any life cycle process.

This process is governed by the following controls and enablers:

- *Applicable Laws and Regulations*
- *Industry Standards* – relevant industry specifications and standards
- *Agreements* – terms and conditions of the agreements
- *Project Procedures and Standards* – including project plans
- *Project Directives*
- *Organization/Enterprise Policies, Procedures, and Standards* – including guidelines and reporting mechanisms
- *Organization/Enterprise Infrastructure*
- *Project Infrastructure*.

5.6.1.4 Outputs

Outputs of the Information Management Process include the following:

- *Information Management Strategy*

- *Information Repository* – Supports the availability for use and communication of all relevant systems artifacts in a timely, complete, valid, and, if required, confidential manner
- *Information Management Report*.

5.6.1.5 Process Activities

The Information Management Process includes the following activities:

- *Plan Information Management*
  - Support establishing and maintaining a system data dictionary – see project planning outputs
  - Define system-relevant information, storage requirements, access privileges, and the duration of maintenance
  - Define formats and media for capture, retention, transmission, and retrieval of information
  - Identify valid sources of information
- *Perform Information Management*
  - Periodically obtain artifacts of information
  - Maintain information according to security and privacy requirements
  - Retrieve and distribute information, as required
  - Archive designated information for compliance with legal, audit, and knowledge retention requirements
  - Retire unwanted, invalid, or unverifiable information according to organizational policy, security, and privacy requirements.

Common approaches and tips:

- In the Project Planning Process (see Section 5.1), an Information Management Plan is tailored to satisfy the individual project procedures for information management. An Information Management Plan identifies the system-relevant information to be collected, retained, secured, and disseminated, with a schedule for retirement.
- Identify information-rich artifacts and store them for later use even if the information is informal, such as a design engineer's notebook.

- Information management delivers value to the organization and the project by using a variety of mechanisms to provide access to the contents of data repositories. Email, web-based access through intranets, and database queries are a few examples.
- ISO 17799, *Code of Practice for Information Security Management*, is an international standard that provides a best practices framework for implementing security controls.
- ISO 10303, *Standard for the Exchange of Product Model Data (STEP),* includes Application Protocol (AP) 239, *Product Life-cycle Support* (PLCS), which addresses information requirements for complex systems.

5.6.2 Elaboration

5.6.2.1 Information Management Concepts

The purpose of Information Management is to maintain an archive of information produced throughout the system's life cycle. The initial planning efforts for information management are defined in the Information Management Plan, which establishes the scope of project information that is maintained; identifies the resources and personnel skill level required; defines the tasks to be performed; and identifies information management tools and processes, as well as methodology, standards, and procedures that will be used on the project. Typical information includes source documents from stakeholders, contracts, project planning documents, verification documentation, engineering analysis reports, and the files maintained by configuration management. Today, information management is most often concerned with the integration of databases, such as the decision database, and the ability to access the results from decision gate reviews and other decisions taken on the project; requirements management tools and databases; computer-based training and electronic interactive user manuals; websites; and shared information spaces over the internet, such as INCOSE Connect. The STEP — ISO 10303 standard provides a neutral computer-interpretable representation of product data throughout the life cycle. ISO 10303-239 (AP239), *Product Life-Cycle Support*, is an international standard that specifies an information model that defines what information can be exchanged and represented to support a product through life.[37] INCOSE is a co-sponsor of ISO 10303-233 (AP233), *SE Data Exchange*. Figure 5-16 shows how AP233 would be used to exchange data between a SysML™ and other SE application and then to applications in the larger life cycle of systems potentially using related ISO STEP data exchange capabilities.

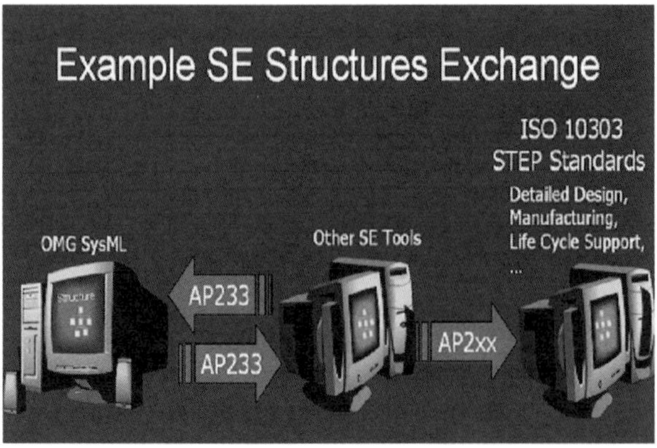

Figure 5-16 AP233 facilitates data exchange

With effective information management, information is readily accessible to authorized project and organization personnel. Challenges related to maintaining databases, security of data, sharing data across multiple platforms and organizations, and transitioning when technology is updated are all handled by information management. With all the emphasis on knowledge management, organizational learning, and information as competitive advantage, these activities are gaining increased attention.

5.7 Measurement Process

5.7.1 Overview

5.7.1.1 Purpose

As stated in ISO/IEC 15288:2008:

> *The purpose of the Measurement Process is to collect, analyze, and report data relating to the products developed and processes implemented within the organization, to support effective management of the processes, and to objectively demonstrate the quality of the products.*[38]

5.7.1.2 Description

The SE Measurement Process helps to define the types of information needed to support program management and technical decisions and implement SE best practices to improve performance. The key SE measurement objective is to measure the SE work products and processes with respect to program/project and organization needs, including timeliness, meeting performance requirements and quality attributes, product conformance to standards, effective use of resources, and continuous process improvement in reducing cost and cycle time.

The guide for *Practical Software and Systems Measurement* (PSM, Section 1.1) states:[39]

> *Measurement provides objective information to help the project manager:*
>
> - *Communicate effectively throughout the project organization*
> - *Identify and correct problems early*
> - *Make key tradeoffs*
> - *Track specific project objectives*
> - *Defend and justify decisions.*

Specific measures are based on information needs and how that information will be used to make decisions and take action. Measurement thus exists as part of a larger management process and includes not just the project manager, but also systems engineers, analysts, designers, developers, integrators, logisticians, etc. The decisions to be made motivate the kinds of information to be generated and, therefore, the measurements to be made.

Another concept of successful measurement is the communication of meaningful information to the decision makers. It is important that the people who use the measurement information understand what is being measured and how it is to be interpreted. Figure 5-17 presents the context diagram for the Measurement Process.

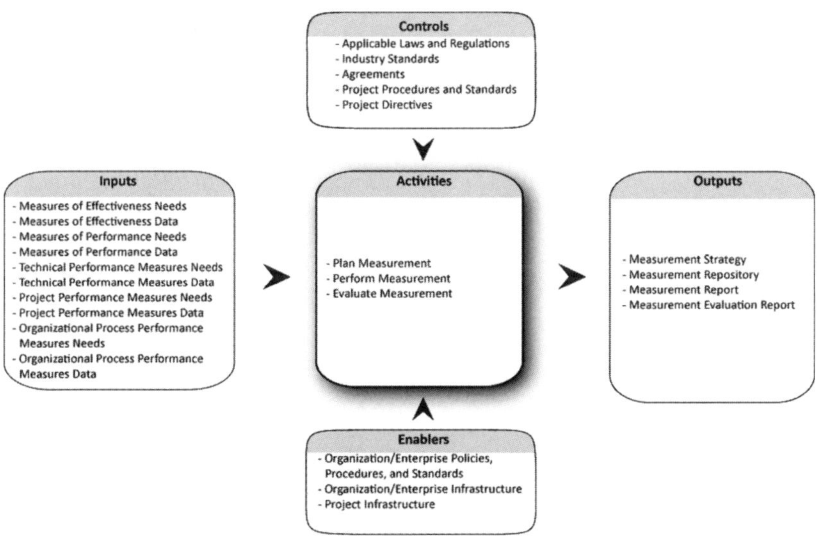

Figure 5-17 Context Diagram for the Measurement Process

5.7.1.3 Inputs

Inputs to the Measurement Process include the following[7]:

- *Measurement Needs* – Identified measurements needs including:
  - MOE Needs
  - MOPs Needs
  - TPM Needs
  - Project Performance Measures Needs
  - Organizational Process Performance Measures Needs
- *Measurement Data* – Data provided for the identified measurement needs including:
  - MOE Data
  - MOP Data
  - TPMs Data
  - Project Performance Measures Data
  - Organizational Process Performance Measures Data.

This process is governed by the following controls and enablers:

- *Applicable Laws and Regulations*
- *Industry Standards* – relevant industry specifications and standards
- *Agreements* – terms and conditions of the agreements
- *Project Procedures and Standards* – including project plans
- *Project Directives*
- *Organization/Enterprise Policies, Procedures, and Standards* – including guidelines and reporting mechanisms
- *Organization/Enterprise Infrastructure*
- *Project Infrastructure.*

5.7.1.4 Outputs

Outputs of the Measurement Process include:

- *Measurement Strategy*

- *Measurement Repository* – Supports the availability for use and communication of all relevant measures in a timely, complete, valid, and, if required, confidential manner.
- *Measurement Report* – Including documentation of the measurement activity results; the measurement data that was collected, analyzed, and results that were communicated; and any improvements or corrective actions driven by the measures with their supporting data.
- *Measurement Evaluation Report* – Including analysis of the Measurement Process and any suggested improvements or necessary corrective actions.

5.7.1.5 Process Activities

The Measurement Process includes the following activities:

- *Plan Measurement*
  - Establish a measurement strategy
  - Identify the measurement stakeholders
  - Identify and prioritize the information needs of the decision makers and stakeholders
  - Identify and select relevant measures that aid with the management and technical performance of the program
  - Define the base measures, derived measures, indicators, data collection, measurement frequency, measurement repository, reporting method and frequency, trigger points or thresholds, and review authority
- *Perform Measurement*
  - Collect, store and verify the data per plan
  - Process and analyze the data to obtain measurement results (information products)
  - Document and review the measurement information products with the measurement stakeholders and recommend action, as warranted by the results
- *Evaluate Measurement*
  - Evaluate the effectiveness of the measures for providing the necessary insight for decisions

- Evaluate the effectiveness, efficiency, and compliance of the Measurement Process
- Assign corrective actions, if required
- Document and store all program measures and corrective actions in a measurement repository.

Common approaches and tips:

- Collection of measures for the sake of collection is a waste of time and effort.
- Each measure collected should be regularly reviewed by the measurement stakeholders. At a minimum, key measures should be reviewed monthly and weekly for the more mature organizations.
- Some contracts identify MOEs that must be met. The derived MOPs and TPMs that provide the necessary insight into meeting the MOEs are automatic measures to be included within the measurement plan. Other measures to consider should provide insight into technical and programmatic execution of the program.[7]
- The best measures require minimal effort to collect and are repeatable, straight forward to understand, and presented in a format on a regular (weekly or monthly) basis with trend data.
- Many methods are available to present the data to the measurement stakeholders. Line graphs and control charts are two of the more frequently used. Tools are available to help with measurement.
- If a need for corrective action is perceived, further investigation into the measures may be necessary to identify the root cause of the issue to ensure that corrective actions address the cause instead of a symptom.
- Measurement by itself does not control or improve process performance. Measurement results must be provided to decision makers in a manner that provides the needed insight for the right decisions to be made.

5.7.2 Elaboration

5.7.2.1 Measurement Concepts

Measurement concepts have been expanded upon in previous works that the SE measurement practitioner should reference for further insights:

1. *Measurement Primer*, March 1998, INCOSE, <http://www.incose.org>

2. *Technical Measurement Guide*, Version 1.0, INCOSE & PSM, December 2005, <http://www.incose.org and http://www.psmsc.com>

3. *Systems Engineering Leading Indicators Guide*, Version 1.0, June 15, 2007, LAI, INCOSE, PSM, and SEARI, <http://www.incose.org> and <http://www.psmsc.com>

4. ISO/IEC 15939, *Systems and Software Engineering - Measurement Process*, ISO/IEC 2007

5. *PSM Guide V4.0c, Practical Software and Systems Measurement*, Department of Defense and U.S. Army, October 2000

6. *CMMI® (Measurement and Quantitative Management Process Areas)*, Version 1.2, August 2006, Software Engineering Institute, <http://www.sei.cmu.edu>

7. *Practical Software Measurement: Objective Information for Decision Makers*, Addison-Wesley, 2002.

5.7.2.2 Leading Indicators

A leading indicator is a measure for evaluating the effectiveness of a how a specific activity is applied on a program in a manner that provides information about impacts that are likely to affect the system performance or SE effectiveness objectives. A leading indicator may be an individual measure, or collection of measures, that are predictive of future system performance before the performance is realized. Leading indicators aid leadership in delivering value to customers and end users while assisting in taking interventions and actions to avoid rework and wasted effort.

Leading indicators differ from conventional SE measures in that conventional measures provide status and historical information, while leading indicators use an approach that draws on trend information to allow for predictive analysis (forward looking). By analyzing the trends, predictions can be forecast on the outcomes of certain activities. Trends are analyzed for insight into both the entity being measured and potential impacts to other entities. This provides leaders with the data they need to make informed decisions and, where necessary, take preventative or corrective action during the program in a proactive manner. While the leading indicators appear similar to existing measures and often use the same base information, the difference lies in how the information is gathered, evaluated, interpreted, and used to provide a forward looking perspective. Examples of leading indicator measures include the following:

- *Requirements Trends* – Rate of maturity of the system definition against the plan. Requirements trends characterize the stability and completeness of the system requirements that could potentially impact design and production.
- *Interface Trends* – Interface specification closure against plan. Lack of timely closure could pose adverse impact to system architecture, design, implementation and/or V&V, any of which could pose technical, cost, and schedule impact.
- *Requirements Validation Trends* – Progress against the plan in ensuring that the customer requirements are valid and properly understood. Adverse trends would pose impacts to system design activity with corresponding impacts to technical, cost, and schedule baselines and customer satisfaction.

For a more detailed treatment of this topic, please consult the *Systems Engineering Leading Indicators Guide* referenced in Section 5.7.2.1.

5.7.2.3 Measures of Effectiveness and Measures of Performance

Measures of effectiveness (MOEs) and measures of performance (MOPs) are two concepts that represent types of measures typically collected. MOEs are defined in the *INCOSE-TP-2003-020-01, Technical Measurement Guide*, as follows:

> *The "operational" measures of success that are closely related to the achievement of the mission or operational objective being evaluated, in the intended operational environment under a specified set of conditions; i.e., how well the solution achieves the intended purpose. (Adapted from DoD 5000.2, DAU, and INCOSE)*[7]

Measures of performance are defined as follows:

> *The measures that characterize physical or functional attributes relating to the system operation, measured or estimated under specified testing and/or operational environment conditions. (Adapted from DoD 5000.2, DAU, INCOSE, and EPI 280-04, LM Integrated Measurement Guidebook)*[7]

Measures of effectiveness, which are stated from the acquirer (customer/user) viewpoint, are the acquirer's key indicators of achieving the mission needs for performance, suitability, and affordability across the life cycle. Although they are independent of any particular solution, MOEs are the overall operational success criteria (e.g., mission performance, safety, operability, operational availability, etc.) to be used by the acquirer for the delivered system, services, and/or processes. INCOSE-TP-2003-020-01, *Technical Measurement Guide*,

Section 3.2.1, provides further discussion on this topic.

Measures of performance measure attributes considered as important to ensure that the system has the capability to achieve operational objectives. MOPs are used to assess whether the system meets design or performance requirements that are necessary to satisfy the MOEs. MOPs should be derived from or provide insight for MOEs or other user needs. INCOSE-TP-2003-020-01, *Technical Measurement Guide*, Section 3.2.2, provides further discussion on this topic.

5.7.2.4 Technical Performance Measures

Technical Performance Measures (TPMs) are defined in INCOSE-TP-2003-020-01, *Technical Measurement Guide*, as follows:

> *TPMs measure attributes of a system element to determine how well a system or system element is satisfying or expected to satisfy a technical requirement or goal.*

Technical Performance Measures are used to assess design progress, compliance to performance requirements, or technical risks and provide visibility into the status of important project technical parameters to enable effective management, thus enhancing the likelihood of achieving the technical objectives of the project. TPMs are derived from or provide insight for the MOPs focusing on the critical technical parameters of specific architectural elements of the system as it is designed and implemented. The relationship between TPMs and MOPs is illustrated in Section 3.2.6 of the *INCOSE Technical Measurement Guide*. Selection of TPMs should be limited to critical technical thresholds or parameters that, if not met, put the project at cost, schedule, or performance risk. The TPMs are not a full listing of the requirements of the system or system element. The SEP should define the approach to TPMs.[7]

Without TPMs, a project manager could fall into the trap of relying on cost and schedule status alone, with perhaps the verbal assurances of technical staff to assess project progress. This can lead to a product developed on schedule and within cost that does not meet all key requirements. Values are established to provide limits that give early indications if a TPM is out of tolerance, as illustrated in Figure 5-18.

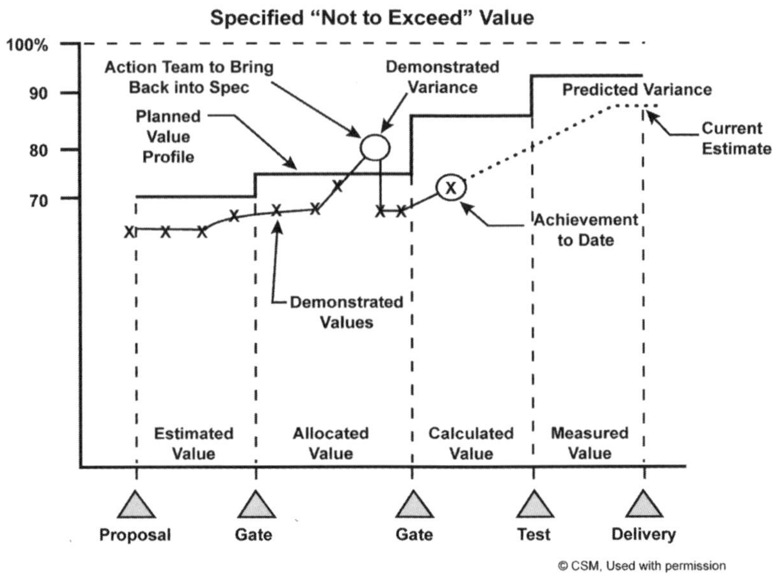

Figure 5-18 TPM Monitoring

Periodic recording of the status of each TPM provides the continuing verification of the degree of anticipated and actual achievement of technical parameters. Measured values that fall outside an established tolerance band alert management to take corrective action. INCOSE-TP-2003-020-01, *Technical Measurement Guide*, Section 3.2.3, provides further discussion on this topic.

5.8 References

1. ISO/IEC 15288:2008, p. 24
2. Adapted from Alexander Kossiakoff and William N. Sweet, *Systems Engineering; Principles and Practice*, Hoboken, NJ: J, Wiley & Sons, Inc., 2003, p. 91
3. ISO/IEC 15288:2008, p. 25
4. Martin, J. N., *Systems Engineering Guidebook*, CRC Press, 1996
5. Forsberg, K., H. Mooz, H. Cotterman, *Visualizing Project Management,* 3rd Ed., J. Wiley & Sons, 2005, p. 198-219
6. see <http://www.pmi.org>
7. Roedler, Garry J., and Cheryl Jones, *Technical Measurement: A Collaborative Project of PSM, INCOSE, and Industry,* INCOSE-TP-2003-020-01, INCOSE Measurement Working Group, 27 December 2006
8. ISO/IEC 15288:2008, p. 27
9. Ibid, p. 29
10. Teale, M., et al., *Management Decision-making,* Prentice Hall, 2003
11. Bayes, the Reverend Thomas (1702 - 1761), "Essay Toward Solving a Problem in the Doctrine of Chances." Philosophical Transactions of the Royal Society of London, 1763
12. Raiffa, Howard, *Decision Analysis: Introductory Lectures on Choices Under Uncertainty*, Addison-Wesley, New York, 1968.

13. Schlaiffer, Robert, *Analysis of Decisions Under Uncertainty*, New York: McGraw-Hill Book Company, 1969.
14. Skinner, David C., *Introduction to Decision Analysis*, 2nd Ed., Gainsville, FL: Probabilistic Publishing, 2001, p. 34, 88
15. Poppendieck, Mary, "Morphing the Mold", Software Development, 7 July 2003
16. Emes, Michael, "Strategic Multi-stakeholder Trade Studies," *Proceedings of EuSEC*, Edinburgh, September 2006, p. 18-20
17. Analytical Hierarchy Process is described at <http://www.expertchoice.com>
18. David Ullman, *Making Robust Decisions*, Trafford Publishing, 2007
19. ISO/IEC 15288:2008, p. 29
20. Conrow, E. H., *Effective Risk Management*, 2nd Ed., American Institute of Aeronautics and Astronautics, Inc., 2003, p 435
21. Ibid. p. 436
22. Fossnes, T., "Lessons from Mt. Everest Applicable to Project Leadership." *Proceedings of the 15th Annual INCOSE International Symposium*, Rochester, NY, 2005
23. see <http://www.theirm.org>
24. Forsberg, K., H. Mooz, H. Cotterman, *Visualizing Project Management*, 3 Ed., J. Wiley & Sons, 2005, p. 223-253
25. ISO/IEC 16085:2006, *Systems and software engineering – Life cycle processes – Risk management*
26. Widerman, R. Max. *Comparative Glossary of Common Project Management Terms*, v3.1, March 2002
27. Project Management Institute. *A Guide to the PMBOK.* Upper Darby, USA, 2000, p. 176
28. *AT&T Engineering Guides for Managing Risk: Design to Reduce Technical Risk; Design's Impact on Logistics; Moving a Design into Production; Testing to Verify Design and Manufacturing Readiness,* from the McGraw-Hill Professional Book Group (1-800-842-3075)
29. Barton, Thomas L., William G. Shenkir, and Paul L. Walker, *Making Enterprise Risk Management Pay Off: How Leading Companies Implement Risk Management*, Upper Saddle River, NJ: Financial Times/Prentice Hall PTR/Pearson Education Company, 2002
30. Michel, Robert Mark, and Dan Galai, *Risk Management*, NY: McGraw Hill, 2001
31. Shaw, T. E. and Lake, J. G., Ph.D., "Systems Engineering: The Critical Product Development Enabler," *APICS Conference*, April 1993
32. Wideman, R. Max (Ed.), *Project and Program Risk Management: A Guide to Managing Project Risks and Opportunities*, Newtown Square, PA: Project Management Institute, 2004
33. This section adapted from Forsberg, K., H. Mooz, H. Cotterman, *Visualizing Project Management,* 3rd Ed., J. Wiley & Sons, 2005
34. ISO/IEC 15288:2008, p. 32
35. Ibid, p. 33
36. *STSC Crosstalk*, <http://www.stsc.hill.af.mil/crosstalk/2003/05/brykczynski.html>
37. see <http://www.plcs-resources.org/>
38. ISO/IEC 15288:2008, p. 34
39. *Practical Software and Systems Measurement: A Foundation for Objective Project Management*, v. 4.0b1, <http://www.psmsc.com/PSMGuide.asp>

This page intentionally left blank.

6 Agreement Processes

The initiation of a project begins with user need. Once a need is perceived and resources are committed to establish a project, it is possible to define the parameters of an acquisition and supply relationship. This relationship exists whenever an organization with a need does not have the ability to satisfy that need without assistance. Agreement processes are defined in ISO/IEC 15288:2008 as follows:

> *[Agreement] processes define the activities necessary to establish an agreement between two organizations. If the Acquisition Process is invoked, it provides the means for conducting business with a supplier: of products that are supplied for use as an operational system, of services in support of operational activities, or of elements of a system being developed by a project. If the Supply Process is invoked, it provides the means for conducting a project in which the result is a product or service that is delivered to the acquirer.*[1]

Acquisition is also an alternative for optimizing investment when a supplier can meet the need in a more economical or timely manner. The Acquisition and Supply Processes are the subject of Sections 6.1 and 6.2, respectively.

Virtually all organizations interface with industry, academia, government, customers, partners, etc. An overall objective of Agreement processes is to identify these external interfaces and establish the parameters of these relationships, including identifying the inputs required from the external entities and the outputs that will be provided to them. This network of relationships provides the context of the business environment of the organization and access to future trends and research. Some relationships are defined by the exchange of products or services.

The acquisition and Supply Processes are two sides of the same coin. Each process establishes the contractual context and constraints under which the other system life-cycle processes are performed. The unique activities for the agreement processes are related to contracts and managing business relationships. An important contribution of ISO/IEC 15288:2008 is the recognition that systems engineers are relevant contributors in this domain.[2] The Maglev train case (see Section 3.6.3) is an example where the government representatives of China and Germany participated in the relationship.

Contract negotiations are handled in various ways depending on the specific organization. In a process that is widely used, the contracts organization in industry (or the contracting officer in the government) is responsible for negotiating contracts, including the contract terms and conditions. Key parameters, such as profit target and acceptable contract type (firm fixed price, cost plus fixed fee, cost plus award fee), are established by the business area manager or by organization management. Project managers rarely lead contract negotiations, however, the lead contract negotiator should only agree to any changes in scope, cost, or schedule with the project manager's approval. The systems engineer is in a supporting role to the project manager during negotiations.

The lead contract negotiator may need, within minutes or a few hours, an assessment of the impact of customer-proposed changes; for major changes, the team may need a few days. In preparation for contract negotiations, systems engineers often perform preliminary trade studies on a range of cost, schedule, and technical performance options that might be proposed by the customer or supplier during negotiations. Of particular importance is the impact to project risk. What is needed is accuracy – not precision – so the team is prepared for anything reasonable that might arise. A team that is prepared will always have a more favorable outcome in negotiations, and the buyer will be pleased to work with a knowledgeable provider.

A critical element to each party is the definition of acceptance criteria, such as:

1. Percent completion of the SRD
2. Requirements stability and growth measures, such as the number of requirements added, modified, or deleted during the preceding time interval (e.g., month, quarter, etc.)
3. Percent completion of each contract requirements document: SOW, RFP, Contract Data/Document Requirements List (CDRL), etc.

These criteria protect both sides of the business relationship – the acquirer from being coerced into accepting a product with poor quality, and the supplier from the unpredictable actions of a fickle or indecisive buyer.

Two Agreement Processes are identified by ISO/IEC 15288:2008: the Acquisition Process and the Supply Process. They are included in this handbook because they conduct the essential business of the organization and establish the relationships between organizations relevant to the acquisition and supply (i.e., buying and selling) of products and services. Agreements may exist between organizational units internal or external to the organization.

6.1 Acquisition Process

6.1.1 Overview

6.1.1.1 Purpose

As stated in ISO/IEC 15288:2008:

> *The purpose of the Acquisition Process is to obtain a product or service in accordance with the acquirer's requirements.*[3]

The Acquisition Process is invoked to establish an agreement between two organizations under which one party acquires products or services from the other. The acquirer experiences a need for an operational system, for services in support of an operational system, for elements of a system being developed by a project, or for services in support of project activities. Often our experience with the Acquisition Process is typified by the purchase of telephones or automobiles. To facilitate the purchase of more complex services and products is a primary SE responsibility. The start of an acquisition/Supply Process begins with the determination of, and agreement on, user needs. The goal is to find a supplier that can meet those needs.

6.1.1.2 Description

The role of the acquirer demands familiarity with the Technical, Project, and Organizational Project-Enabling Processes as it is through them that the supplier will execute the agreement. An acquirer organization applies due diligence in the selection of a supplier to avoid costly failures and impacts to the organization budgets and schedules. This section is written from the perspective of the acquirer organization. Figure 6-1 is the context diagram for the Acquisition Process.

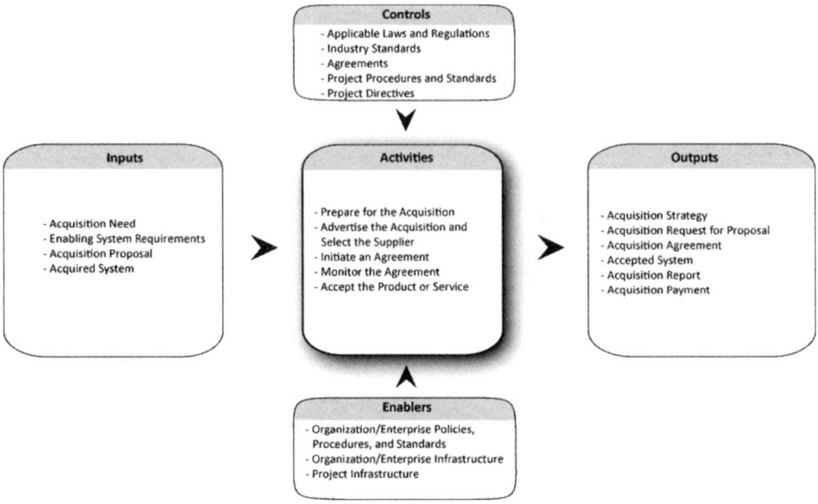

Figure 6-1 Acquisition Process Context Diagram

6.1.1.3 Inputs

The inputs to the Acquisition Process include the following:

- *Acquisition Need* – The Acquisition Process begins with the identification of a need that cannot be met within the organization encountering the need, or a need that can be met in a more economical way by a supplier.
- *Enabling System Requirements* – Necessary systems that enable the realization of the system-of-interest are acquired, including:
  - Implementation Enabling System Requirements
  - Integration Enabling System Requirements
  - Verification Enabling System Requirements
  - Transition Enabling System Requirements
  - Validation Enabling System Requirements
  - Operation Enabling System Requirements
  - Maintenance Enabling System Requirements
  - Disposal Enabling System Requirements.
- *Acquisition Proposal* – The responses of one or more candidate suppliers in response to an acquisition RFP.

- *Acquired System* – The system, subsystem, or system element (product or service) is delivered to the acquirer consistent with the delivery conditions of agreement.

This process is governed by the following controls and enablers:

- *Applicable Laws and Regulations*
- *Industry Standards* – relevant industry specifications and standards
- *Agreements* – terms and conditions of the agreements
- *Project Procedures and Standards* – including project plans
- *Project Directives*
- *Organization/Enterprise Policies, Procedures, and Standards* – including guidelines and reporting mechanisms
- *Organization/Enterprise Infrastructure*
- *Project Infrastructure*.

It is important to note that availability of adequate funding is essential to beginning the Acquisition Process.

6.1.1.4 Outputs

Outputs of the Acquisition Process include the following:

- *Acquisition Strategy* – May also include inputs to determine acquisition constraints.
- *Acquisition Request for Proposal (RFP)* – The organization identifies candidate suppliers that could meet this need, and the acquisition personnel identify a plan for procuring the system-of-interest. Inputs are received from the project management and engineering personnel in the organization with the need.
- *Acquisition Agreement* – The agreement can vary from formal contracts or less formal inter-organizational work orders.
- *Accepted System* – Responsibility for system-of-interest is transferred from supplier to acquirer and the product or service is available to the project.
- *Acquisition Report*
- *Acquisition Payment* – Payments or other compensations is rendered.

6.1.1.5 Process Activities

The Acquisition Process includes the following activities:

- *Prepare for the Acquisition*
  - Develop and maintain Acquisition Plans, Strategies, Policies, Procedures to meet the organization goals and objectives and the needs of the project management and technical SE organizations.
  - Identify a list of potential suppliers – suppliers may be internal or external to the acquirer organization.
- *Advertise the Acquisition and Select the Supplier*
  - Identify needs in an RFP – Through the use of the Technical Processes, including Requirements Analysis, the acquiring organization produces a set of requirements that will form the basis for the procurement specification.
  - Select appropriate suppliers – Using selection criteria, rank suppliers by their suitability to meet the overall need and establish supplier preferences and corresponding justifications. Viable suppliers should be willing to conduct ethical negotiations, able to meet technical obligations, and willing to maintain open communications throughout the Acquisition Process.
  - Evaluate supplier responses to the RFP – Ensure the system-of-interest meets acquirer needs and complies with industry and other standards. Assessments from the Project Portfolio Management and Quality Management Processes and recommendations from the requesting organization are necessary to determine the suitability of each response and the ability of the supplier to meet the stated commitments. Record recommendations from evaluation of responses to the RFP. This can range from formal documentation to less formal inter-organizational interactions (e.g., between design engineering and marketing).
  - Select the preferred supplier based on acquisition criteria.
- *Initiate an Agreement*
  - Negotiate agreement – Acquirer commits to specify requirements for system-of-interest; participate in verification, validation, and acceptance activities; render payment according to the schedule; participate in exception and change control

procedures; and contribute to transparent risk management procedures. The agreement will establish criteria for assessing progress toward final delivery.

- Establish delivery acceptance criteria – The procurement specification, in the context of the overall agreement, should clearly state the criteria by which the acquirer will accept delivery from the supplier. A verification matrix can be used to clarify these criteria.

- *Monitor the Agreement*
  - Manage Acquisition Process activities, including decision-making for agreements, relationship building and maintenance, interaction with organization management, responsibility for the development of plans and schedules, and final approval authority for deliveries accepted from the supplier.
  - Maintain communications with supplier, stakeholders, and other organizations regarding the project.
  - Assess execution of agreements to identify risks and issues, progress towards mitigation of risks, adequacy of progress toward delivery, and cost and schedule performance, and to determine potential undesirable outcomes for the organization.
  - Amend agreements when impacts on schedule, budget, or performance are identified.
- *Accept the Product or Service*
  - Accept delivery of products and services – in accordance with all agreements and relevant laws and regulations.
  - Render payment – or other agreed consideration in accordance with agreed payment schedules.
  - Accept responsibility in accordance with all agreements and relevant laws and regulations.
  - When an Acquisition Process cycle concludes, a final review of performance is conducted to extract lessons-learned for continued process performance.

Common approaches and tips:

- Establish acquisition guidance and procedures that inform acquisition planning, including recommended milestones, standards, assessment criteria and decision gates. Include approaches for identifying, evaluating, choosing, negotiating, managing and terminating suppliers.

- Establish a point of responsibility within the organization for monitoring and controlling individual agreements. This person maintains communication with the supplier and is part of the decision-making team to assess progress in the execution of the agreement. The possibility of late delivery or cost overruns should be identified and communicated to the organization as early as noted.
- Define and track measures that indicate progress on agreements. Appropriate measures require the development of tailored measures that do not drive unnecessary and costly efforts but do provide the information needed to ensure the progress is satisfactory and that key issues and problems are identified early to allow time for resolution with minimal impact to the delivery and quality of the product and service.
- Include technical representation in the selection of the suppliers to critically assess the capability of the supplier to perform the required task. This helps reduce the risk of contract failure and its associated costs, delivery delays, and increased resource commitment needs. Past performance is highly important, but changes to key personnel should be identified and evaluated.
- Communicate clearly with the supplier about the real needs and avoid conflicting statements or making frequent changes in the statement of need that introduce risk into the process.
- Maintain traceability between the supplier's responses to the acquirer's solicitation. This can reduce the risk of contract modifications, cancellations, or follow-on contracts to fix the product or service.
- Institute for supply management has useful guidance for purchasing and marketing.[4]

6.1.2 Elaboration

Elaborations for this section are under development and will be included in a future revision of the *INCOSE Systems Engineering Handbook*.

6.2 Supply Process

6.2.1 Overview

6.2.1.1 Purpose

As stated in ISO/IEC 15288:2008:

> *The purpose of the Supply Process is to provide an acquirer with a product or service that meets agreed requirements.*[5]

The Supply Process is invoked to establish an agreement between two organizations under which one party supplies products or services to the other. Within the supplier organization, a project is conducted according to the recommendations of this handbook with the objective of providing a product or service that meets the contracted requirements. In the case of a mass produced product or service, a marketing function may represent the acquirer and establish customer expectations.

6.2.1.2 Description

The Supply Process is highly dependent upon the Technical, Project, and Organizational Project-Enabling Processes as it is through them that the work of executing the agreement is accomplished. This means that the Supply Process is the larger context in which the other processes are applied under contract. This section is written from the perspective of the supplier organization. Figure 6-2 is the context diagram for the Supply Process.

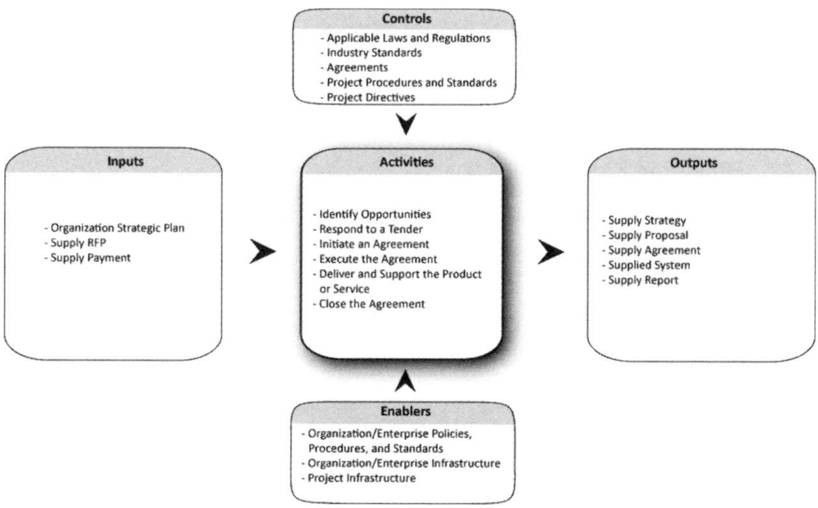

Figure 6-2 Supply Process Context Diagram

6.2.1.3 Inputs

The inputs to the Supply Process include the following:

- *Organization Strategic Plan* – Supply decisions are determined within in the context of the overall organizational strategy.
- *Supply Request for Proposal (RFP)* – A request from another organization to propose a solution to meet a need for a system (product or service).
- *Supply Payment* – Payments or other compensations is received and acknowledged.

This process is governed by the following controls and enablers:

- *Applicable Laws and Regulations*
- *Industry Standards* – relevant industry specifications and standards
- *Agreements* – terms and conditions of the agreements
- *Project Procedures and Standards* – including project plans
- *Project Directives*
- *Organization/Enterprise Policies, Procedures, and Standards* – including guidelines and reporting mechanisms
- *Organization/Enterprise Infrastructure*
- *Project Infrastructure.*

6.2.1.4 Outputs

Outputs of the Supply Process include the following:

- *Supply Strategy* – May also include inputs to determine supply constraints. Should also include the identification of potential acquirers.
- *Supply Proposal* – The organization responds to the supply RFP.
- *Supply Agreement* – The agreement can vary from formal contracts or less formal inter-organizational work orders.
- *Supplied System* – The system-of-interest (product or service) is delivered according to delivery conditions of agreement.
- *Supply Report.*

6.2.1.5 Process Activities

The Supply Process includes the following activities:

- *Identify Opportunities*
  - Develop and maintain Supply plans, strategies, policies, procedures to meet the organization goals and objectives and the needs of the project management and technical SE organizations.
- *Respond to a Tender*
  - Select appropriate acquirers willing to conduct ethical negotiations, able to meet financial obligations, and willing to maintain open communications throughout the Supply Process.
  - Evaluate acquirer requests and prepare a response – Ensure satisfactory response proposes a system-of-interest that meets acquirer needs and complies with industry and other standards. Assessments from the Project Portfolio Management, Human Resource Management, and Quality Management Processes are necessary to determine the suitability of this response and the ability of the organization to meet these commitments.
- *Initiate an Agreement*
  - Negotiate agreement – Supplier commits to meet requirements for system-of-interest; meet delivery milestones, verification, validation, and acceptance conditions; accept payment schedule; execute exception and change control procedures; and maintain transparent risk management procedures. The agreement will establish criteria for assessing progress toward final delivery.
- *Execute the Agreement*
  - Execute the agreement – Start a project and invoke the other processes defined in this handbook.
  - Manage Supply Process activities, including decision-making for agreements, relationship building and maintenance, interaction with organization management, responsibility for the development of plans and schedules, and final approval authority for deliveries made to acquirer.
  - Maintain communications with acquirer, sub-suppliers, stakeholders, and other organizations regarding the project.

  - Assess execution of agreements to identify risks and issues, progress towards mitigation of risks, and adequacy of progress toward delivery; evaluate cost and schedule performance; and determine potential undesirable outcomes for the organization.

- *Deliver and Support the Product or Service*
  - Deliver products and services in accordance with all agreements and relevant laws and regulations.
- *Close the Agreement*
  - Receive and acknowledge payment or other agreed consideration in accordance with agreed payment schedules.
  - Transfer responsibility in accordance with all agreements and relevant laws and regulations.
  - When a Supply Process cycle concludes, a final review of performance is conducted to extract lessons-learned for continued process performance.

Common approaches and tips:

- Agreements fall into a large range from formal to very informal based on verbal understanding. Contracts may call for a fixed price, cost plus fixed fee, incentives for early delivery, penalties for late deliveries, and other financial motivators.
- Relationship building and trust between the parties is a non-quantifiable quality that, while not a substitute for good processes, makes the human interactions agreeable.
- Develop technology white papers or similar documents to demonstrate and describe to the (potential) acquirer the range of capabilities in areas of interest. Use traditional marketing approaches to encourage acquisition of mass produced products.
- Maintain an up-to-date internet presence, even if the organization does not engage in electronic commerce.
- When expertise is not available within the organization (e.g. legal and other governmental regulations, laws, etc.), retain subject matter experts to provide information and specify requirements related to agreements.
- Invest sufficient time and effort into understanding acquirer needs before the agreement. This can improve the estimations for cost and schedule and positively affect agreement execution. Evaluate any

technical specifications for the product or service for clarity, completeness and consistency.

- Involve personnel who will be responsible for agreement execution to participate in the evaluation of and response to the acquirer's request. This reduces the start-up time once the project is initiated, which in turn is one way to recapture the cost of writing the response.
- Make a critical assessment of the ability of the organization to execute the agreement; otherwise, the high risk of failure and its associated costs, delivery delays, and increased resource commitment needs will reflect negatively on the reputation of the entire organization.

6.2.2 Elaboration

Elaborations for this section are under development and will be included in a future revision of the *INCOSE Systems Engineering Handbook*.

6.3 References

1. ISO/IEC 15288:2008, p. 15
2. Arnold, Stuart, and Harold Lawson, "Viewing Systems from a Business Management Perspective: The ISO/IEC 15288 Standard", *Journal of Systems Engineering,* 7:3, 2003
3. ISO/IEC 15288:2008, p. 15
4. see <http://www.napm.org/>
5. ISO/IEC 15288:2008, p. 16

This page intentionally left blank.

7 Organizational Project-Enabling Processes

Organizational Project-Enabling Processes are the purview of the organization (also known as enterprise) and are used to direct, enable, control, and support the system life cycle. Organizational Project-Enabling Processes are defined in ISO/IEC 15288:2008 as follows:

> *The Organizational Project-Enabling Processes ensure the organization's capability to acquire and supply products or services through the initiation, support and control of projects. They provide resources and infrastructure necessary to support projects and ensure the satisfaction of organizational objectives and established agreements. They are not intended to be a comprehensive set of business processes that enable strategic management of the organization's business.*[1]

This chapter focuses on the capabilities of an organization relevant to the realization of a system; as stated above, they are not intended to address general business management objectives, although sometimes the two overlap.

Organizational units cooperate to develop, implement, deploy, operate, maintain and dispose of the system-of-interest. Enabling systems may also need to be modified to meet the needs of new systems, developed or acquired if they do not exist. Examples include development, manufacturing, training, verification, transport, maintenance, and disposal systems that support the system-of-interest.

Five Organizational Project-Enabling Processes are identified by ISO/IEC 15288:2008. They are: Life Cycle Model Management, Infrastructure Management, Product Portfolio Management, Human Resource Management, and Quality Management. The organization will tailor these processes and their interfaces to meet specific strategic and communications objectives. Figure 1-1, System Life Cycle Processes Overview, depicts how Organizational Project-Enabling Processes impact the Technical, Project, and Agreement processes described in this handbook.

7.1 Life Cycle Model Management Process

7.1.1 Overview

7.1.1.1 Purpose

As stated in ISO/IEC 15288:2008:

> *The purpose of the Life Cycle Model Management Process is to define, maintain, and assure availability of policies, life cycle processes, life cycle models, and procedures for use by the organization with respect to the scope of [ISO/IEC 15288:2008].*
>
> *This process provides life cycle policies, processes, models, and procedures that are consistent with the organization's objectives, that are defined, adapted, improved and maintained to support individual project needs within the context of the organization, and that are capable of being applied using effective, proven methods and tools.*[2]

The resulting life cycle model is a "framework of processes and activities concerned with the life cycle that may be organized into stages, which also acts as a common reference for communication and understanding."[3]

The value propositions to be achieved by instituting organization-wide processes for use by projects are as follows:

1. Provide repeatable/predictable performance across the projects in the organization (this helps the organization in planning and estimating future projects and in demonstrating reliability to customers)
2. Leverage practices that have been proven successful by certain projects and instill those in other projects across the organization (where applicable)
3. Enable process improvement across the organization
4. Improve ability to efficiently transfer staff across projects as roles are defined and performed consistently
5. Improve start up of new projects (less re-inventing the wheel).

In addition, the standardization across projects may enable cost savings through economies of scale for support activities (tool support, process documentation, etc).

7.1.1.2 Description

This process (1) establishes and maintains a set of policies and procedures at the organization level that support the organization's ability to acquire and supply products and services, and (2) provides integrated system life cycle models necessary to meet the organization's strategic plans, policies, goals, and objectives for all projects and all system life-cycle stages. The processes are defined, adapted, and maintained to support the requirements of the organization, SE organizational units, individual projects, and personnel. Life Cycle Model Management Processes are supplemented by recommended methods and tools. The resulting guidelines in the form of organization policies and procedures are still subject to tailoring by projects, as discussed in Chapter 8. Figure 7-1 is the context diagram for this process.

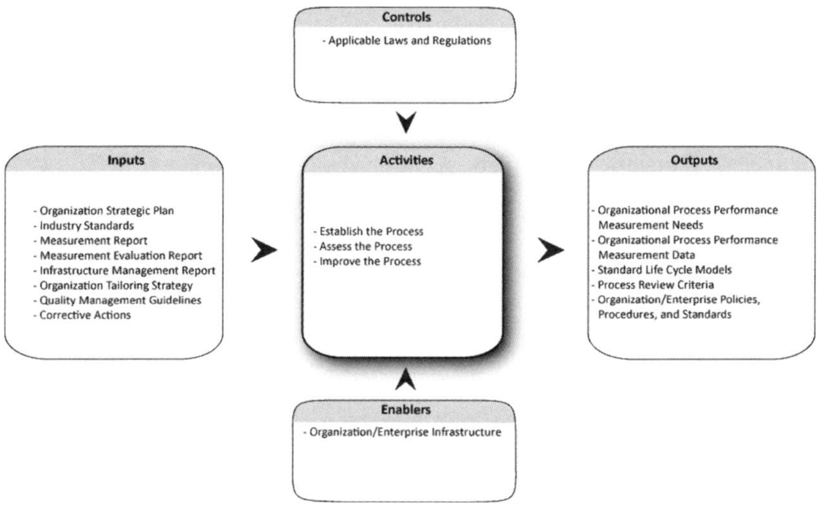

Figure 7-1 Life-Cycle Model Management Process Context Diagram

7.1.1.3 Inputs

Inputs to the Life Cycle Model Management Process include the following:

- *Organization Strategic Plan* – Organization strategic plans and infrastructure are used to ensure consistency in the eventual recommendations.
- *Industry Standards* – This handbook and relevant standards, new knowledge from research, and industry sponsored knowledge networks are examples of the sources from which Life Cycle Model Management Processes are extracted.

- *Measurement Report* – Assessments from projects and trends collected from Tailoring Processes provide constructive input for improvements to an organization's life cycle model implementation.
- *Measurement Evaluation Report* – Assessments of the Measurement Process provide constructive input for improvements to an organization's life cycle model implementation.
- *Infrastructure Management Report* – Reports from the Infrastructure Management Process provide constructive input for improvements to an organization's life cycle model implementation.
- *Organization Tailoring Strategy* – Relating to incorporating new or updated external standards.
- *Quality Management Guidelines*
- *Corrective Actions* – Resulting from process-related reviews and audits.

This process is governed by the following controls and enablers:

- *Applicable Laws and Regulations*
- *Organization/Enterprise Infrastructure*.

7.1.1.4 Outputs

Outputs of the Life Cycle Model Management Process include the following:

- *Organizational Process Performance Measures Needs* – Identification of measurement criteria indicating the degree to which the system life cycle processes are being followed
- *Organizational Process Performance Measures Data* – Data provided to measure the degree to which the system life cycle processes are being followed
- *Standard Life Cycle Models* – Including definition of the business and other decision-making criteria regarding entering and exiting each life cycle stage
- *Process Review Criteria* – Including criteria for assessments and approvals/disapprovals
- *Organization/Enterprise Policies, Procedures, and Standards* – Organization process guidelines in the form of organization policies, directives, and procedures for applying the system life cycle processes and adapting them to meet the needs of individual projects (e.g., templates for management plans, such as configuration, information,

and RMPs). This also includes defining accountability and authority for all SE and management processes within the organization and the roles and responsibilities for SE and management processes within the organization.

7.1.1.5 Process Activities

The Life Cycle Model Management Process includes the following activities:

- *Establish the Process*
  - Identify sources (organization, corporate, industry, academia, stakeholders and customers) of Life Cycle Model Management Process information
  - Distill the information from multiple sources into an appropriate set of life cycle models that are aligned with the organization and business area plans and infrastructure
  - Establish Life Cycle Model Management guidelines in the form of plans, policies, procedures, tailoring guidance, models, and methods and tools for controlling and directing the life cycle models
  - Define, integrate, and communicate life cycle model roles, responsibilities, authorities, requirements, measures, and performance criteria based on the Life Cycle Model Management Process guidelines
  - Define the decision-making criteria that determine entering and exiting each stage of the system life cycle – expressed in terms of business achievements
  - Disseminate policies, procedures, and directives throughout the organization.
- *Assess the Process*
  - Conduct periodic reviews of the life cycle models used by projects – use assessments to confirm the adequacy and effectiveness of the Life Cycle Model Management Processes
- *Improve the Process*
  - Identify opportunities to improve the organization Life Cycle Model Management guidelines on a continuing basis based on individual project assessments, individual feedback, and changes in the organization strategic plan

- Communicate with all relevant organizations regarding the creation of and changes in the Life Cycle Model Management guideline.

Common approaches and tips:

- Base policies and procedures on an organization-level strategic and business area plan that provides a comprehensive understanding of the organization's goals, objectives, stakeholders, competitors, future business, and technology trends.
- Ensure that policy and procedure compliance review is included as part of the business decision gate criteria.
- Development of a Life Cycle Model Management Processes intranet and information database with essential information provides an effective mechanism for disseminating consistent guidelines and providing announcements about organization-related topics, as well as industry trends, research findings, and other relevant information. This provides a single point of contact for continuous communication regarding the Life Cycle Model Management guidelines and encourages the collection of valuable feedback and the identification of organization trends.
- Establish an organization center of excellence for Life Cycle Model Management Processes. This organization can become the focal point for the collection of relevant information, dissemination of guidelines, and analysis of assessments and feedback. They can also develop checklists and other templates to support project assessments to ensure that the pre-defined measures and criteria are used for evaluation.
- Manage the network of external relationships by assigning personnel to identify standards, industry and academia research, and other sources of organization management information and concepts needed by the organization.

  The network of relationships includes government, industry, and academia. Each of these external interfaces provide unique and essential information for the organization to succeed in business and meet the continued need and demand for improved and effective systems and products for its customers. It is up to the Life Cycle Model Management Process to fully define and utilize these external entities and interfaces (i.e., their value, importance, and capabilities that are required by the organization):

  - Legislative, regulatory, and other government requirements
  - Industry SE and management related standards, training, capability maturity models
  - Academic education, research results, future concepts and perspectives, requests for financial support.
- Establish an organization communication plan for the policies and procedures. Most of the processes in this handbook include dissemination activities. An effective set of communication methods is needed to ensure that all stakeholders are well-informed.
- Methods and tools for enabling the application of Life Cycle Model Management Processes must be effective and tailored to the implementation approach of the organization and its projects. A responsible organization can be created or designated to coordinate the identification and development of partnerships and/or relationships with tool vendors and working groups. They can recommend the use of methods and tools that are intended to help personnel avoid confusion, frustration, and wasting valuable time and money. These experts may also establish an integrated tool environment between interacting tools to avoid cumbersome (and inaccurate) data transfer.
- Including stakeholders, such as engineering and project management organizations, as participants in developing the Life Cycle Model Management guidelines increases their commitment to the recommendations and incorporates a valuable source of organization experience.
- Developing alternative life cycle models based on the type, scope, complexity, and risk of a project decreases the need for tailoring by engineering and project organizations.
- Provide clear guidelines for tailoring and adaptation.
- Work continually to improve the life cycle models and processes.

7.1.2 Elaboration

7.1.2.1 Standard SE Processes

An organization engaged in SE provides the requirements for establishing, maintaining, and improving the standard SE process and the policies, practices, and supporting functional processes (see Figure 7-2) necessary to meet customer needs throughout the organization. Further, it defines the process for tailoring the standard SE process for use on projects and for making improvements to the project-tailored SE processes.

Organizational management must review and approve the standard SE process and changes to it. Organizations should consider establishing an SE Process Group (SYSPG) to oversee SE process definition and implementation.

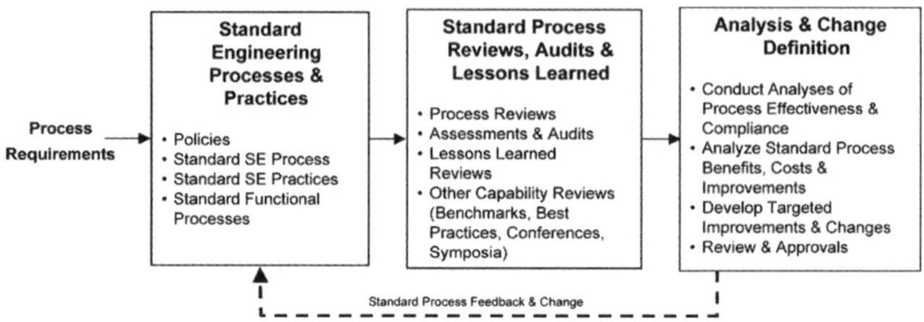

Figure 7-2 Standard SE Process Flow

An organization establishes a standard SE process using a reference SE process model, which is tailored by projects to meet specific customer and stakeholder needs. The reference model should tailor industry, government, or other agency "best practices" based on multiple government, industry, and organization reference SE process documents. The reference SE model must include an SE improvement process. Projects are expected to follow this process, as tailored to meet project-specific SE process needs. The standard process must be tailorable, extensible, and scalable to meet a diverse range of projects, from small study contracts to large projects requiring thousands of participants.

The standard SE process model is established by selection of specific processes and practices from this handbook, industry SE process references (such as ANSI/EIA-632 and ISO/IEC 15288:2008), and government SE process references, as appropriate and is applicable to every engineering capability maturity focus area or process area in the Capability Maturity Model® Integration (CMMI®) approach.

A high-performing organization also reviews the process (as well as work products), conducts assessments and audits (e.g., CMMI® assessments and ISO audits), retains corporate memory through the understanding of lessons learned, and establishes how benchmarked processes and practices of related organizations can affect the organization. Successful organizations should analyze their process performance, its effectiveness and compliance to organizational and higher directed standards, and the associated benefits and costs, and then develop targeted improvements.

The basic requirements for standard and project-tailored SE process control, based on CMMI®, are as follows:

1. SE processes shall be identified for use on projects.
2. Implementation and maintenance of SE processes shall be documented.
3. Inputs and outputs shall be defined for SE subprocesses.
4. Entrance and exit criteria shall be defined for SE process major activities.
5. Projects shall use a defined set of standard methods or techniques in the SE process.
6. Tailoring guidelines shall be used to permit the standard process to meet project-specific needs.
7. Project management shall identify what parts of the standard SE process have been tailored to meet project-specific needs.
8. Strengths and weaknesses in the SE process shall be assessed.
9. The SE process shall be periodically assessed.
10. The SE process shall be compared to benchmark processes used by other organizations.

In addition, basic requirements specifically for SE process improvement control from these standards are as follows:

1. Organization best practices shall be identified and communicated to projects.
2. The standard SE process shall identify areas for future improvement.
3. SE process users shall be able to identify proposed improvements.
4. Compliance with improvement processes, plans, and practices shall be verified.
5. The project-tailored SE improvement process shall include a means for evaluating its effectiveness.
6. The project-tailored SE improvement process shall include a means for making needed improvements.
7. The standard SE process work products shall be reviewed and results used to improve the process.

8. The standard SE process compliance shall be reviewed and results used to improve the process.

7.1.2.2 Reviews, Audits, Lessons Learned, and Analysis for Change

The standard SE process must meet requirements for review, assessment, and audit, and for establishing lessons learned and best practices.

Reviews

Organizations that set internal policies and procedures conduct periodic Process Compliance Reviews (PCR) to assess the effectiveness, strengths, and weaknesses of their processes. Such reviews are conducted on a recurring basis, as determined by the SE organization with management involvement, and may be combined with other assessments (such as ISO 9000) to reduce the perceived burden. The review should address defects in the examined process at the time of the review, as well as the improvement process, tailoring of the process, and tailoring of the improvement process (if applicable). A standard SE process checklist should be used as the basis for this PCR. It may be augmented by additional issues and questions. Each review may address a subset of the standard SE checklist. The questions asked and results from this review should be recorded and stored. Organization management prioritizes and approves changes based on requested or recommended areas for improvement in the standard processes.

The PCR must be organized by a PCR Coordinator who will notify responsible personnel of the specific dates, formats, and requirements for the reviews; define the lists of required attendees and invitees; and set the agenda. The data presented in these reviews should be archived. Key results from PCRs must be provided for management consideration. The PCR should cover at least the following:

- Identify strengths and weaknesses in the SE process and its improvement process.
- Identify key process elements that need to be followed in large and/or small projects
- Identify areas for future improvement
- Address the effectiveness of the tailored improvement process
- Address the conduct of, defects in, and improvements to the SE improvement process
- Review SE work products to identify potential trends indicating possible systemic issues

- Review the results of PCRs to identify potential trends indicating possible systemic issues
- Review a sampling of in-process reviews to identify potential trends indicating possible systemic issues
- Review the definition and use of SE process measures.

Occasional benchmark comparisons with other organizations can be useful. Reference processes, practices, and other capabilities can be accessed through either direct contact or an intermediary's compilations of benchmarked processes, practices, and other capabilities.

Leaders in management research advise organizations also to reassess the utility of process management programs and apply them with discrimination, but not without a caution:[4]

> *In the appropriate setting, process management activities can help companies improve efficiency, but the risk is that you misapply these programs, in particular in areas where people are supposed to be innovative. Brand new technologies to produce products that don't exist are difficult to measure. This kind of innovation may be crowded out when you focus too much on processes you can measure.*[5]

Audits

Audits and assessments should be conducted to include internal and external assessments of capability maturity, and internal and external audits of key SE processes and the personnel that implement them. One practice is to ensure that the auditors are from outside of the direct project management reporting chain, to facilitate objectivity in the audit and to ensure that the auditor is comfortable that they can report project findings without negative consequences from their management.

Internal assessments of capability maturity should be conducted to improve the organization's SE process and to prepare for external assessments. The assessment team should consist of at least one external, qualified lead assessor. The standard for use in capability assessments should be an external, industry formal standard, such as CMMI® or equivalent.

External assessments of capability maturity should be conducted when justified by the business needs of the organization. They should be led by an external, qualified lead assessor, with a significant part of the assessment team consisting of external, qualified assessors. The standard for use in capability assessments should be the external, industry formal standard required by the organization or customer, such as CMMI® or equivalent.

Internal audits of organizations using SE processes should be conducted to prepare for an external audit of the organization. The audit should investigate defects (i.e. process errors) in the SE process in use at the time of the audit to understand why the defects or errors occurred. As with the PCR, a standard SE process activity checklist should be used as the basis for this audit but may be augmented by additional issues and questions. Each audit may address a subset of the standard checklist. The questions asked and results from this audit must be recorded and stored.

Lessons Learned

Lessons learned are needed to recognize an organization's good and best practices, understand the lessons of the organization's history, and avoid repeating the mistakes of the past. They must address technical SE management and engineering, specialty management and engineering, and any other project or organization activities affecting SE processes.

The SE organization must plan and follow through to collect lessons learned at pre-defined milestones in the system life cycle. The SE organization should periodically review lessons learned to analyze and improve SE processes and practices. It should also establish best practices and capture them in an easy-to-retrieve form.

To obtain the benefits in capturing lessons learned, it is important that the organization create a lessons learned repository in which to pool lessons learned from various projects and to disseminate the information to new and on-going projects. This repository will need to be communicated to new projects, with the ability to filter to relevant types of lessons learned categories, and also build in processes to review lessons learned during critical junctures in projects (such as project planning, decision making).

Analysis and Change

The SYSPG should sample and monitor project implementation of tailored SE process activities to provide insight into project SE process effectiveness and compliance and understand project issues and concerns. These analyses should identify and define potential process strengths, weaknesses, deficiencies, and problem areas in the standard SE process that are revealed in the project process assessments. Feedback, minutes, and reports from project assessments, audits, formal reviews, in-process reviews, and PCRs should also be sampled and analyzed, as should the results of training evaluations, action item compliance reports, lessons learned reports, and best practices. Analyses should address at least the following issues:

1. Is the SE process effective and useful (e.g., are we getting what we need from it)?

2. Can the SE process be improved (e.g., are there process elements that were a "waste of time", or that should have been or could have been done better)?
3. What can we change in the SE process to make it better (e.g., what could we do to eliminate the recorded action items or defects)?
4. What is the productivity of the standard major SE process elements?
5. Are the SE support tools and facilities effective and useful?
6. Is information being collected on the effectiveness and usefulness of the SE process?
7. Is information being used to improve the effectiveness and usefulness of the SE process?

These analyses will not evaluate or judge project performance; rather, they will focus on internal standard SE process improvement and establish a basis for SE process:

- Effectiveness
- Utility
- Utility of support tools and facilities
- Issues and concerns
- Compliance in the organization
- Understanding of implementation impacts
- Potential systemic problems
- Potential for improvement.

Rationales for and results from decisions should be recorded and stored.

7.1.2.3 Continuous Process Improvement

The organization should improve the SE process based in large part on usage, experience, and feedback from the programs. The standard SE process improvement should be managed and improved with the participation and support of key stakeholders in the process activities.

Changes may be required, requested, or recommended based on prioritized areas for improvement, PCRs, and/or exception reports. The SYSPG should prioritize the requested or recommended areas for improvement, identify the specific changes needed, assess the impact of consequential changes, and recommend adjustments that can be made in the standard SE process to implement the priority improvements within budget and schedule. Assessments must consider trends in process technology, such as changes in capability maturity assessment practices wanted by organizational stakeholders. Areas of improvement should be developed from the results of project work, product reviews, management reviews, and SE PCRs. If the improvements or changes involve errors or problems with the existing process, these are identified to determine the actions needed to prevent future occurrences. Management should approve the prioritized areas for improvement, decide on what changes will be made, and adjust budgets and labor estimates, as needed, to enable the changes to be accomplished.

The SYSPG should identify and refine the criteria being used in analyses and assessments to improve their focus on essential organization business and process needs. Criteria should be recorded and stored. The SYSPG should also assess the areas for improvement and related analyses to determine if additional tailoring guidelines are needed. If so, they should identify the tailoring changes needed, fit them into the overall improvement priority scheme, and recommend which changes should be made. As described above, analyses and assessments determine whether changes are needed in the standard SE process and its documentation, including changes to tailoring guidance to better meet project-specific needs.

An SE Process Improvement Plan should be developed and updated at least annually based in part on targeted improvements and results from reviews. After the SYSPG have prepared standard SE process changes, they will be submitted to SE and project management for approval, with the coordination of the project managers. The SYSPG should document and store process compliance in process compliance and/or exception reports.

7.2 Infrastructure Management Process

7.2.1 Overview

7.2.1.1 Purpose

As stated in ISO/IEC 15288:2008:

> *The purpose of the Infrastructure Management Process is to provide the enabling infrastructure and services to projects to support organization and project objectives throughout the life cycle.*
>
> *This process defines, provides and maintains the facilities, tools, and communications and information technology assets needed for the organization's business with respect to the scope of [ISO/IEC 15288:2008].*[6]

7.2.1.2 Description

The work of the organization is accomplished through projects, which are conducted within the context of the infrastructure environment. This infrastructure needs to be defined and understood within the organization and the project to ensure alignment of the working units and achievement of overall organization strategic objectives. This process exists to establish, communicate, and continuously improve the system life-cycle process environment. Figure 7-3 contains the context diagram for the Infrastructure Management Process.

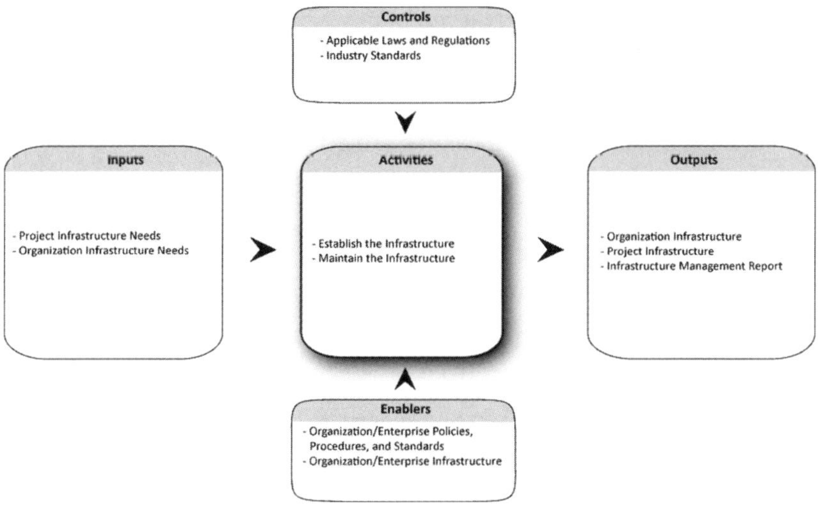

Figure 7-3 Infrastructure Management Process Context Diagram

7.2.1.3 Inputs

The inputs to the Infrastructure Management Process include the following:

- *Project Infrastructure Needs* – Specific requests for infrastructure products or services from the projects, including commitments to external stakeholders
- *Organization Infrastructure Needs* – Specific requests for infrastructure products or services from the organization, including commitments to external stakeholders.

This process is governed by:

- *Applicable Laws and Regulations*
- *Industry Standards* – relevant industry specifications and standards

- *Organization/Enterprise Policies, Procedures, and Standards* – including guidelines and reporting mechanisms
- *Organization/Enterprise Infrastructure*.

7.2.1.4 Outputs

Outputs of the Infrastructure Management Process include the following:

- *Organization/Enterprise Infrastructure* – Resources and services that support the organization.
- *Project Infrastructure* – Resources and services that support a project.
- *Infrastructure Management Report* – including cost, usage, downtime/response measures, etc. These can be used to support capacity planning for upcoming projects.

7.2.1.5 Process Activities

The Infrastructure Management Process includes the following activities:

- *Establish the Infrastructure*
  - Gather and negotiate infrastructure resource needs with organization and projects
  - Identify, obtain, and provide the infrastructure resources and services
- *Maintain the Infrastructure*
  - Manage infrastructure resource availability to ensure organization goals and objectives are met. Conflicts and resource shortfalls are managed with steps for resolution.
  - Allocate infrastructure resources and services to support all projects
  - Control multi-project infrastructure resource management communications to effectively allocate resources throughout the organization, identify potential future or existing conflict issues and problems with recommendations for resolution.

Common approaches and tips:

- Qualified resources may be leased (in-sourced or out-sourced) in accordance with the investment strategy.

- Establish an organization infrastructure architecture. Integrating the infrastructure of the organization can make the execution of routine business activities more efficient.
- Establish a resource management information system with enabling support systems and services to maintain, track, allocate and improve the resources for present and future organization needs. Computer-based equipment tracking, facilities allocation, and other systems are recommended for organizations over 50 people.
- Attend to physical factors, including facilities and human factors, such as ambient noise level and computer access to specific tools and applications.
- Begin planning in early life-cycle stages of all system development efforts to address utilization and support resource requirements for system transition, facilities, infrastructure, information/data storage, and management. Enabling resources should also be identified and integrated into the organization's infrastructure.

7.2.2 Elaboration

7.2.2.1 Infrastructure Management Concepts

Projects all need resources to meet their objectives. Project planners determine the resources needed by the project and attempt to anticipate both current and future needs. The Infrastructure Management Process provides the mechanisms whereby the organization infrastructure is made aware of project needs and the resources are scheduled to be in place when requested. While this can be simply stated, it is less simply executed. Conflicts must be negotiated and resolved, equipment must be purchased and sometimes repaired, buildings need to be refurbished, and information technology services are in a state of constant change. The infrastructure management organization collects the needs, negotiates to remove conflicts, and is responsible for providing the enabling organization infrastructure without which nothing else can be accomplished. Since resources are not free, their costs are also factored into investment decisions. Financial resources are addressed under the Project Portfolio Management Process, but all other resources, except for human resources (see Section 7.4), are addressed under this process.

Infrastructure Management is complicated by the number of sources for requests, the need to balance the skills of the labor pool against the other infrastructure elements (e.g., computer-based tools), the need to maintain a balance between the budgets of individual projects and the cost of resources, the need to keep apprised of new or modified policies and procedures that might influence the skills inventory, and myriad unknowns.

Resources are allocated based on requests. Infrastructure management collects the needs of all the projects in the active portfolio and schedules or acquires non-human assets, as needed. Additionally, the Infrastructure Management Process maintains and manages the facilities, hardware, and support tools required by the portfolio of organization projects. Infrastructure management is the efficient and effective deployment of an organization's resources when and where they are needed. Such resources may include inventory, production resources, or information technology. The goal is to provide materials and services to a project when they are needed to keep the project on target and on budget. An optimized goal is to achieve 100% utilization of every resource, but this is unlikely when providing some minimum level of service while minimizing cost. Infrastructure management relies heavily on forecasts into the future of the demand and supply of various resources.

The organization environment and subsequent investment decisions are built on the existing organization/enterprise infrastructure, including facilities, equipment, personnel, and knowledge. Efficient use of these resources is achieved by exploiting opportunities to share enabling systems or to use a common system element on more than one project. These opportunities are enabled by good communications within the organization. Integration and interoperability of supporting systems, such as financial, human resources (see Section 7.4), and training, is critically important to executing organization strategic objectives. Feedback from active projects is used to refine and continuously improve the infrastructure.

Further, trends in the market may suggest changes in the supporting environment. The availability and suitability of the organization infrastructure and associated resources is one of the critical project assessments and provides feedback for improvement–and reward–mechanisms. All organization processes require mandatory compliance with government and corporate laws and regulations. Decision-making is governed by the organization strategic plan.

7.3 Project Portfolio Management Process

7.3.1 Overview

7.3.1.1 Purpose

As stated in ISO/IEC 15288:2008:

> *The purpose of the Project Portfolio Management Process is to initiate and sustain necessary, sufficient and suitable projects in order to meet the strategic objectives of the organization.*

This process commits the investment of adequate organization funding and resources, and sanctions the authorities needed to establish selected projects. It performs continued qualification of projects to confirm they justify, or can be redirected to justify, continued investment.

NOTE: This process is applied within the system context. The projects in question are focused on the systems-of-interest for the organization.[7]

Project Portfolio Management also provides organizational output to external stakeholders, such as parent organizations, investors/funding sources, and governance bodies.

7.3.1.2 Description

Projects create the products or services that generate income for an organization. Thus, the conduct of successful projects requires an adequate allocation of funding and resources and the authority to deploy them to meet project objectives. Most business entities manage the commitment of financial resources using well defined and closely monitored processes.

The Project Portfolio Management Process also performs ongoing evaluation of the projects in its portfolio. Based on periodic assessments, projects are determined to justify continued investment if they have the following characteristics:

- Progress toward achieving established goals
- Comply with project directives from the organization
- Are conducted according to an approved plan
- Provide a service or product that is still needed and providing acceptable investment returns.

Otherwise, projects may be redirected, or in extreme instances, cancelled. Figure 7-4 shows the context diagram for the Project Portfolio Management Process.

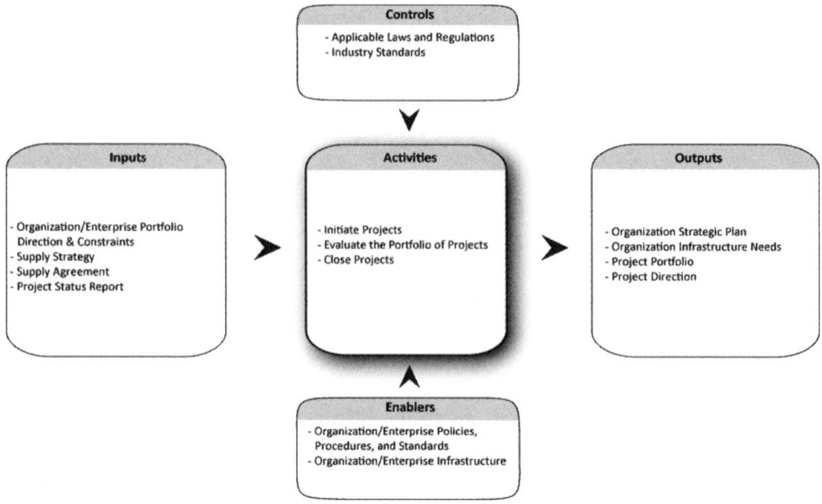

Figure 7-4 Project Portfolio Management Context Diagram

7.3.1.3 Inputs

Inputs to the Project Portfolio Management Process include the following:

- *Organization/Enterprise Portfolio Direction & Constraints* – Includes organization business objectives, funding outlay and constraints, on-going R&D, etc.
- *Supply Strategy* – The portfolio of projects should be consistent with the overall strategy for supplying systems, which includes the identification candidate projects for management consideration.
- *Supply Agreement* – Specific agreement terms may influence or prioritize project initiation and termination decisions.
- *Project Status Report* – Includes status on meeting the objectives set out for the project.

This process is governed by the following controls and enablers:

- *Applicable Laws and Regulations*
- *Industry Standards* – relevant industry specifications and standards
- *Project Procedures and Standards* – including project plans
- *Organization/Enterprise Policies, Procedures, and Standards* – including guidelines and reporting mechanisms
- *Organization/Enterprise Infrastructure.*

7.3.1.4 Outputs

Outputs of the Project Portfolio Management Process include the following:

- *Organization Strategic Plan*
- *Organization Infrastructure Needs* – Specific requests for infrastructure products or services from the organization, including commitments to external stakeholders.
- *Project Portfolio* – Qualified investment opportunities result in the initiation of new project(s), includes resources and budgets identified and allocated to the projects, and clearly defined project management accountability and authorities.
- *Project Direction* – Includes sustainment of projects meeting assessment criteria and redirection or termination of project not meeting assessment criteria.

7.3.1.5 Process Activities

The Project Portfolio Management Process includes the following activities:

- *Initiate Projects*
  - Identify and assess investment opportunities consistent with the organization strategic plan
  - Establish Business Area Plans – use the strategic objectives to identify candidate projects to fulfill them
  - Establish project scope, define project management accountabilities and authorities, identify expected project outcomes
  - Allocate adequate funding and other resources
  - Identify interfaces and opportunities for multi-project synergies
  - Specify project reporting requirements and review schedules that govern the progress of the project
  - Authorize project execution
- *Evaluate the Portfolio of Projects*
  - Evaluate ongoing projects to provide rationale for continuation, redirection, or termination
- *Close Projects*
  - Cancel or suspend projects that are completed or designated for termination.

Common approaches and tips:

- The process of developing the business area plans helps the organization assess where it needs to focus resources to meet present and future strategic objectives. Include representatives from relevant stakeholders in the organization community.
- When investment opportunities present themselves, prioritize them based on measurable criteria such that projects can be objectively evaluated against a threshold of acceptable performance.
- Expected project outcomes should be based on clearly defined, measurable criteria to ensure that an objective assessment of progress can be determined. Specify the investment information that will be assessed for each milestone. Initiation should be a formal milestone that does not occur until all resources are in place as identified in the project plan.
- Establish a program office or other coordination organization to manage the synergies between active projects in the organization portfolio. Complex and large organization architectures require the management and coordination of multiple interfaces and make additional demands on investment decisions. These interactions occur within and between the projects.
- Include risk assessments (see Section 5.4) in the evaluation of ongoing projects. Projects that contain risks that may pose a challenge in the future might require redirection. Cancel or suspend projects whose disadvantages or risks to the organization outweigh the investment.
- Include opportunity assessments in the evaluation of ongoing projects. Addressing project challenges may represent a positive investment opportunity for the organization. Avoid pursuing opportunities that are inconsistent with the capabilities of the organization and its strategic goals and objectives or contain unacceptably high technical risk, resource demands, or uncertainty.
- Allocate resources based on the requirements of the projects; otherwise, the risk of cost and schedule overruns may have a negative impact on quality and performance of the project.
- Establish effective governance processes that directly support investment decision-making and communications with project management.

7.3.2 Elaboration

7.3.2.1 Define the Business Case and Develop Business Area Plans

Project Portfolio Management balances the use of financial assets within the organization. Organization management generally demands that there be some beneficial return for the effort expended in pursuing a project. The business case and associated business area plans establish the scope of required resources (e.g., people and money) and schedule, and set reasonable expectations. An important element of each design gate is a realistic review of the business case as the project matures. The result is re-verification or perhaps restatement of the business case. The Iridium case, described in Section 3.2, illustrates the dangers of failing to keep a realistic perspective. Similarly, despite the technological triumph of implementing the world's first Maglev train line, the exorbitant initial cost and slow return on investment are causing authorities to question plans to build another line.

The business case may be validated in a variety of ways. For large projects, sophisticated engineering model, or even prototypes of key system elements, help prove that the objectives of the business case can be met and that the system will work as envisioned prior to committing large amounts of resources to full-scale engineering and manufacturing development. For very complex systems, such a demonstration can be conducted at perhaps twenty percent of development cost. For smaller projects, when the total investment is modest, proof-of-concept models may be constructed during the Concept Stage to prove the validity of business case assumptions.

Investment opportunities are not all equal, and organizations are limited in the number of projects that can be conducted concurrently. Further, some investments are not well aligned with the overall strategic plan of the organization. For these reasons, opportunities are evaluated against the portfolio of existing agreements and ongoing projects, taking into consideration the attainability of the stakeholders' requirements.

7.4 Human Resource Management Process

7.4.1 Overview

7.4.1.1 Purpose

As stated in ISO/IEC 15288:2008:

> *The purpose of the Human Resource Management Process is to ensure the organization is provided with necessary human resources and to maintain their competencies, consistent with business needs.*

This process provides a supply of skilled and experienced personnel qualified to perform life cycle processes to achieve organization, project and customer objectives.[8]

7.4.1.2 Description

Projects all need resources to meet their objectives. This process deals with human resources. Non-human resources, including tools, databases, communication systems, financial systems and information technology, are addressed using the Infrastructure Management Process (see Section 7.2)

Project planners determine the resources needed for the project by anticipating both current and future needs. The Human Resource Management Process provides the mechanisms whereby the organization management is made aware of project needs and personnel are scheduled to be in place when requested. While this can be simply stated, it is less simply executed. Conflicts must be resolved, personnel must be trained, and employees are entitled to vacations and time away from the job.

The human resource management organization collects the needs, negotiates to remove conflicts, and is responsible for providing the personnel without which nothing else can be accomplished. Since qualified personnel are not free, their costs are also factored into investment decisions. Figure 7-5 is the context diagram for the Human Resources Management Process.

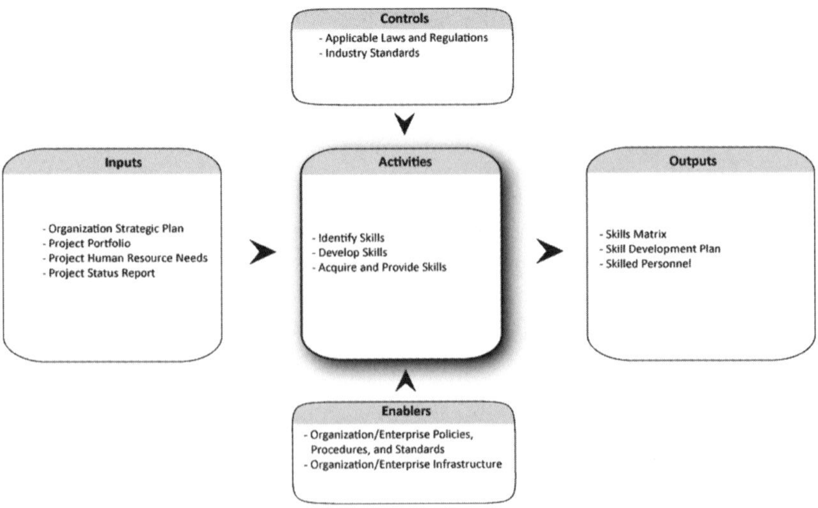

Figure 7-5 Human Resource Management Process Context Diagram

7.4.1.3 Inputs

The inputs to the Human Resource Management Process include the following:

- *Organization Strategic Plan* – Decision-making is governed by the organization strategic plan.
- *Project Portfolio* – Contains the necessary information for all of the organizations projects.
- *Project Human Resource Needs* – The needs of all the projects in the active portfolio are collected.
- *Project Status Report* – Includes status on personnel availability and effectiveness for the project.

This process is governed by the following controls and enablers:

- *Applicable Laws and Regulations*
- *Industry Standards* – relevant industry specifications and standards
- *Project Procedures and Standards* – including project plans
- *Organization/Enterprise Policies, Procedures, and Standards* – including guidelines and reporting mechanisms
- *Organization/Enterprise Infrastructure.*

7.4.1.4 Outputs

Outputs of the Human Resource Management Process include the following:

- *Skills Matrix* – Identify the skill needs of the organization and projects and record the skill of personnel.
- *Skill Development Plan* – Includes the organizational training plan needed to develop internal personnel and the acquisition of external personnel.
- *Skilled Personnel* – The right people with the right skills are assigned to projects per their skill needs and timing.

7.4.1.5 Processes Activities

The Human Resource Management Process includes the following activities:

- *Identify Skills*
  - Manage personnel availability to ensure organization goals and objectives are met. Conflicts and resource shortfalls are managed with steps for resolution.

  - Personnel needs are evaluated against available people with the pre-requisite skills to determine if training or hiring activities are indicated.
- *Develop Skills*
  - Obtain (or develop) and deliver training to close identified gaps
  - Keep employees motivated – content in their career progression, current with their training, and appropriately allocated using techniques that are within acceptable organization and corporate guidelines and constraints
- *Acquire and Provide Skills*
  - Provide human resources to support all projects
  - Obtain qualified personnel when deficits gaps identify that needs cannot be met with existing personnel
  - Control multi-project communications to effectively allocate resources throughout the organization, and identify potential future or existing conflict issues and problems with recommendations for resolution
  - Other related assets are scheduled, or if necessary, acquired.

Common approaches and tips:

- The availability and suitability of personnel is one of the critical project assessments and provides feedback for improvement and reward mechanisms.
- Consider using an IPDT environment as a means to reduce the frequency of project rotation; recognize progress and accomplishments and reward success; and establish apprentice and mentoring programs for newly hired employees and students.
- Maintain a pipeline of qualified candidates that are interested in joining the organization as employees or temporary staff. Focus recruitment, training, and retention efforts on personnel with experience levels, skills, and subject matter expertise demanded by the projects. Personnel assessments should review proficiency, motivation, and ability to work in a team environment, as well as the need to be retrained, reassigned, or relocated.
- Personnel are allocated based on requests and conflicts are negotiated. The goal is to provide personnel to a project when they are needed to keep the project on target and on budget.

- A key concern is keeping project personnel from becoming over-committed, especially persons with specialized skills.
- Skills inventory and career development plans are important documentation that can be validated by engineering and project management.
- Qualified personnel and other resources may be hired temporarily, – in-sourced, or out-sourced in accordance with the organizational strategy.
- Encourage personnel to engage in external networks as a means of keeping abreast of new ideas and attracting new talent to the organization.
- Maintain an organization career development program that is not sidetracked by project demands. Develop a policy that all personnel receive training or educational benefits on a regular cycle. This includes both undergraduate and graduate studies, in-house training courses, certifications, tutorials, workshops, and conferences.
- Remember to provide training on organization policies and procedures and system life cycle processes.
- Establish a resource management information infrastructure with enabling support systems and services to maintain, track, allocate and improve the resources for present and future organization needs. Computer-based human resource allocation and other systems are recommended for organizations over 50 people.
- Use the slack time in the beginning of a project to obtain and train the necessary people to avoid a shortfall of skilled engineers, technologists, managers, and operations experts.
- Trends in the market may suggest changes in the composition of project teams and the supporting IT environment.
- All organization processes require mandatory compliance with government and corporate laws and regulations.
- ISO/IEC 15288:2008 also includes performing knowledge management as part of this process. Consistent with a knowledge management strategy, capture and maintain common and domain information for sharing across the organization.
- Employee performance reviews should be conducted regularly, and career development plans should be managed and aligned to the objectives of both the employee and the organization. Career

development plans should be reviewed, tracked, and refined to provide a mechanism to help manage the employee's career within the organization.

7.4.2 Elaboration

7.4.2.1 Human Resource Management Concepts

The Human Resource Management Process maintains and manages the people required by the portfolio of organization projects. Human resource management is the efficient and effective deployment of qualified personnel when and where they are needed. An optimized goal is to achieve 100% utilization of every resource, but this is unlikely when providing some minimum level of service while minimizing cost. Human resource management relies heavily on forecasts into the future of the demand and supply of various resources.

The primary objective of this process is to provide a pool of qualified personnel to the organization. This is complicated by the number of sources for requests, the need to balance the skills of the labor pool against the other infrastructure elements (e.g., computer-based tools), the need to maintain a balance between the budgets of individual projects and the cost of resources, the need to keep apprised of new or modified policies and procedures that might influence the skills inventory, and myriad unknowns.

Project managers face their resource challenges competing for scarce talent in the larger organization pool. They must balance access to the experts they need for special studies with stability in the project team with its tacit knowledge and project memory. Today's projects depend on teamwork and optimally multi-disciplinary teams. Such teams are able to resolve project issues quickly through direct communication between team members. Such intra-team communication shortens the decision-making cycle and is more likely to result in improved decisions because the multi-disciplinary perspectives are captured early in the process. However, studies have shown that group decisions are often "riskier" resulting in the potential for greater innovation.

In a multi-disciplinary team, each member comes from a discipline with its own perspective. They are responsible for representing that viewpoint, while at the same time establishing the necessary relationships with the other members. However, team results are condemned to mediocrity unless each member confronts the team with challenging ideas while focusing on the final result. Using concurrent development (see Figure 7-6), it may be possible to finish the work faster and thus return valuable skills back to the personnel pool, having achieved a successful delivery.

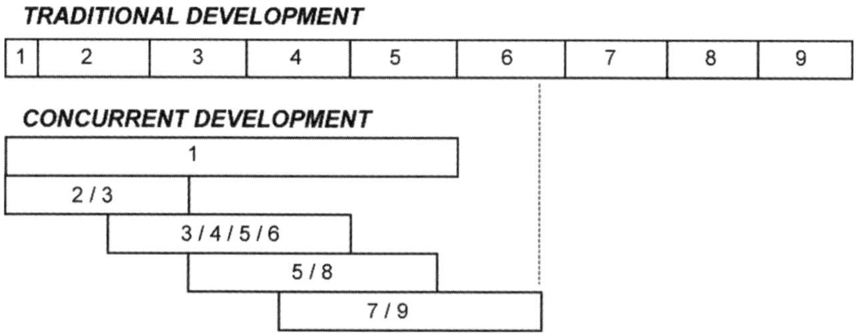

Figure 7-6 Shorter delivery time with concurrent development vs. traditional

As early as 1974, Wilton P. Chase had identified the importance of communications and that "system designs are dependent upon the effective integration of multidisciplinary efforts."[9] He recommended that the organization of a system project should provide opportunity for all disciplinary specialists to work together continuously on a face-to-face basis and, most importantly, to acquire the systems viewpoint and understanding of the role that their specific knowledge can provide in deriving a particular system design.[10]

Chase advocates identifying the lines of communication among tasks in terms of interdependencies and mutual constraints to reveal that different stages of the life cycle call for different tasks and different personnel skills. Properly used, this allows management to acquire and utilize the proper combination of specialist and generalist skills. A project avoids "bureaucratization" of the design approach by streamlining the organization and integrating the various specialist backgrounds into common system-oriented task groups with loyalties directed toward the systems design effort.

Modern projects use the concepts of IPDTs to establish a project organization (see Section 5.1.2.3).[11]

7.5 Quality Management Process

7.5.1 Overview

7.5.1.1 Purpose

As stated in ISO/IEC 15288:2008:

> *The purpose of the Quality Management Process is to assure that products, services and implementations of life cycle processes meet organization quality objectives and achieve customer satisfaction.*[12]

7.5.1.2 Description

The Quality Management Process makes visible the goals of the organization toward customer satisfaction. Since primary drivers in any project are time, cost, and quality, inclusion of a Quality Management Process is essential to every organization. Many of the system life cycle processes are concerned with quality issues, and this forms some of the justification for exerting time, money, and energy into establishing these processes in the organization. Application of this handbook is one approach toward inserting a quality discipline into an organization.

The Quality Management Process establishes, implements, and continuously improves the focus on customer satisfaction and organization goals and objectives. There is a cost to managing quality as well as a benefit. The effort and time required to manage quality should not exceed the overall value gained from the process. Figure 7-7 is the context diagram for the Quality Management Process.

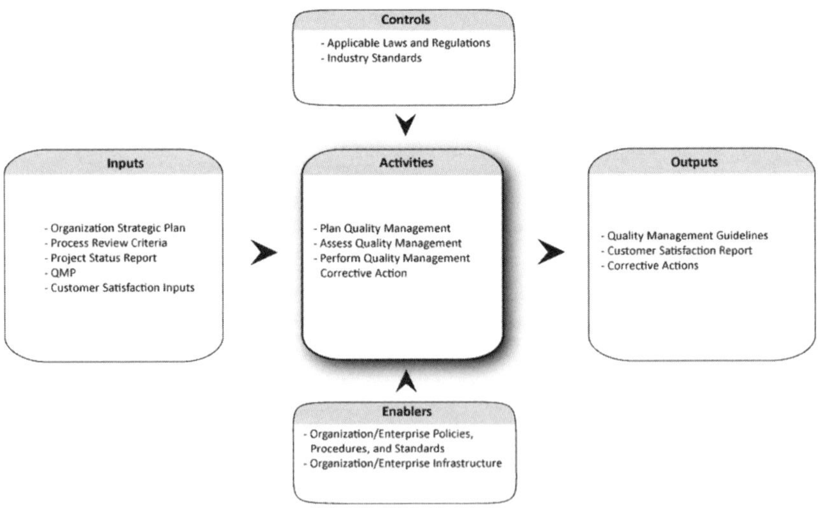

Figure 7-7 Quality Management Process Context Diagram

7.5.1.3 Inputs

Inputs to the Quality Management Process include the following:

- *Organization Strategic Plan* – Decision-making is governed by the organization strategic plan.
- *Process Review Criteria*
- *Quality Management Plan* – The set of project quality plans form the basis of the periodic reviews and audits.

- *Project Status Report* – Includes status on project tailoring an execution.
- *Customer Satisfaction Inputs.*

This process is governed by the following controls and enablers:

- *Applicable Laws and Regulations*
- *Industry Standards* – relevant industry specifications and standards
- *Organization/Enterprise Policies, Procedures, and Standards* – including guidelines and reporting mechanisms
- *Organization/Enterprise Infrastructure.*

7.5.1.4 Outputs

Outputs of the Quality Management Process include the following:

- *Quality Management Guidelines* – This include the set of policies and procedures that apply to quality practices within the organization, within individual projects, and as part of the execution of system life cycle processes and define goals. It also includes quality objectives for processes and systems that are measurable and objective along with assigned accountability and authority for quality management within the organization.
- *Customer satisfaction Report* – Customer satisfaction is closely monitored and any issues addressed.
- *Corrective Actions* – Appropriate actions are taken when quality goals are not achieved.

7.5.1.5 Process Activities

The Quality Management Process includes the following activities:

- *Plan Quality Management*
  - Establish Quality Management Guidelines – Policies, Standards, and Procedures
  - Establish organization and project quality management goals and objectives
  - Define responsibilities and authorities
- *Assess Quality Management*
  - Evaluate project assessments

 - Assess customer satisfaction – against compliance with requirements and objectives
 - Continuously improve the Quality Management Guidelines
- *Perform Quality Management Corrective Action*
 - Recommend appropriate action, when indicated
 - Maintain open communications – within the organization and with stakeholders.

Common approaches and tips:

- Quality is a daily focus – not an afterthought!
- Strategic documentation, including quality policy, mission, strategies, goals, and objectives, provide essential inputs for analysis and synthesis of quality impacts, requirements, and solutions. Existing agreements also provide direction regarding the appropriate level of attention given to quality within the organization.
- Management commitment to quality is reflected in the strategic planning of the organization – the rest of the organization will follow. Everyone in the organization should know the quality policy.
- Development of a Quality Management intranet and information database with essential information provides an effective mechanism for disseminating consistent guidelines, and providing announcements about related topics, as well as industry trends, research findings, and other relevant information. This provides a single point of contact for continuous communication regarding the Quality Management Guidelines and encourages the collection of valuable feedback and the identification of organization trends.
- Analyze statistics from process audits, tests and evaluations, product discrepancy reports, customer satisfaction monitoring, accident and incident reporting, and the implementation of changes to items of a product (e.g., recalled product and/or production lines).
- Quality Management is big business, and a plethora of standards, methods, and techniques exist to help an organization. A short list includes the ISO 9000 series, Total Quality Management (TQM), and Six Sigma (statistical process control). According to ISO 9000, quality is the "Ability of a set of inherent characteristics of a product, system, or process to fulfill requirements of customers and other interested parties."[13]

- A successful strategy is to aim at achieving customer satisfaction primarily by preventing non-fulfillment of requirements. Ideally, customer satisfaction is linked to compliance with requirements. Two indicators that the process is not working are situations where (1) the project is compliant but the customer is unhappy, or (2) the project is not compliant and the customer is happy.
- The consistent involvement and commitment of top management with timely decision-making is mandatory for the quality program. This is reflected in staffing and training of project auditors.
- Project assessments include measurements that can be evaluated to determine the performance of a project team and the progress toward a quality outcome.
- Trends in tailoring of project-specific quality plans provide clear indications of potential improvements in the overall organization guidelines.
- The team of people working in this process will also find a wealth of material in ISO standards and other sources.
- The quality program should have an escalation mechanism so that findings/issues not addressed by the project can be raised to senior management. Projects are under schedule/resource constraints and may not always be responsive to quality findings. In these situations, there should be a mechanism to raise this to senior management to alert them to potential impacts and help them to determine how to proceed.

7.5.2 Elaboration

7.5.2.1 Quality Management Concepts

The purpose of Quality Management is to outline the policies and procedures necessary to improve and control the various processes within the organization that ultimately lead to improved business performance.

The primary objective of QA is to produce an end result that meets or exceeds stakeholder expectations. For example, using a quality system program, manufacturers establish requirements for each type or family of product to achieve products that are safe and effective. To meet this objective, they establish methods and procedures to design, produce, distribute, service, and document devices that meet the quality system requirements. Quality management is closely related to the V&V processes.

Quality Assurance is generally associated with activities such as failure testing, statistical control, and total quality control. Many organizations use statistical process control as a means to achieve Six Sigma levels of quality. Traditional statistical process controls use random sampling to test a fraction of the output for variances within critical tolerances. When these are found, the manufacturing processes are corrected before more bad parts can be produced.

Quality experts[14,15] have determined that if quality cannot be measured, it cannot be systematically improved. Assessment provides the feedback needed to monitor performance, make mid-course corrections, diagnose difficulties, and pinpoint improvement opportunities. A widely used paradigm for QA management is the Plan-Do-Check-Act approach, also known as the Shewhart cycle.[16]

Quality pioneer W. Edwards Deming stressed that meeting user needs represents the defining criterion for quality and that all members of an organization need to participate actively in "constant and continuous" quality improvement—to commit to the idea that "good enough isn't."[17] His advice marked a shift from inspecting for quality after production to building concern for quality into organization processes. As an example, in 1981, Ford launched a quality campaign (see Figure 7-8) that went beyond getting good workers and supporting them with high-quality training, facilities, equipment, and raw materials. By characterizing quality as a "job," everyone in the organization was motivated to concern themselves with quality and its improvement—for every product and customer.[18]

Figure 7-8 Banner from Ford quality campaign

Total Quality Control deals with understanding what the stakeholder/customer really wants. If the original needs statement does not reflect the relevant quality requirements, then quality can be neither inspected nor manufactured into the product. For instance, the Øresund Bridge consortium included not only the bridge material and dimensions but operating, environmental, safety, reliability, and maintainability requirements.

Product certification is the process of certifying that a certain product has passed performance or QA tests or qualification requirements stipulated in regulations, such as a building code or nationally accredited test standards, or that it complies with a set of regulations governing quality or minimum performance requirements. Today, medical device manufacturers are advised to use good judgment when developing their quality system and apply those sections of the Food and Drug Administration Quality System Regulation that are applicable to their specific products and operations. The regulation, 21-CFR-820.5 is continuously updated since its release in 1996. As such, it ought not to be possible to repeat the errors of the Therac-25 project (see Section 3.6.1).

7.6 References

1. ISO/IEC 15288:2008, p. 18
2. Ibid
3. Ibid, p. 4
4. Womack, J. P., "Lean Thinking: Where Have We Been and Where Are We Going?," *Lean Supplement*, Sp 2002, Downloaded May 23, 2006, from <http://www.sme.org/gmn/mag/2002/02sem172/02sem172.html>
5. Brenner, M. J., "TQM, ISO 9000, Six Sigma: Do Process Management Programs Discourage Innovation?," Downloaded May 23, 2006 from <http://knowledge.wharton.upenn.edu/index.cfm?fa=printArticle&ID=1321>
6. ISO/IEC 15288:2008, p. 19
7. Ibid, p. 20
8. Ibid, p. 22
9. Chase, Wilton P., *Management of Systems Engineering*. NY: J. Wiley & Sons, Inc., 1974, p. 14
10. Ibid. p. 22
11. Martin, J. N., *Systems Engineering Guidebook: a process for developing systems and products*, CRC Press, 2000
12. ISO/IEC 15288:2008, p. 23
13. see <http://www.iso.ch/iso/en/iso9000-14000>
14. Crosby, Philip B., *Quality is Free*, New York: New American Library, 1979
15. Juran, J. M. (Ed.), *Quality Control Handbook*, 3rd Ed., New York: McGraw-Hill, 1974
16. Shewhart, Walter Andrew, *Statistical Method from the Viewpoint of Quality Control*, NY: Dover, 1939
17. Deming, W. E., *Out of the Crisis*, MIT Center for Advanced Engineering Study, Cambridge, MA, 1986
18. Scholtes, Peter R., *The Team Handbook: How to Use Teams to Improve Quality*, Joiner Associates, Inc., 1988

This page intentionally left blank.

8 Tailoring Processes

Standards and handbooks are written to address generic practices that may or may not apply to a given organization or system-of-interest. Most are accompanied by a recommendation to adapt the processes and activities to the situation at hand. This adaptation is called tailoring.

Throughout this handbook, advice has appeared about the formal use of SE processes. However, formality is highly dependent on the sophistication of the system, the organizations and the work to be accomplished. Oppressive overhead, with no visible value-added contributions, is demoralizing, and may result in a system that costs more than it is worth. Similarly, insufficient process results in uncoordinated human effort and thrashing[1] – which also adds cost.

The principle behind tailoring is to establish an acceptable amount of process overhead committed to activities not otherwise directly related to the creation of the system. Tailoring scales the rigorous application of SE processes to an appropriate level based on need and the system life-cycle stage. For example, tighter assessment and control cycles are typical of earlier stages of the system life cycle. This chapter describes the process of tailoring SE processes to meet organization and project needs.

8.1 Tailoring Process

8.1.1 Overview

8.1.1.1 Purpose

At the organization level, the Tailoring Process adapts external standards in the context of the organizational processes. At the project level, the Tailoring Process adapts organizational processes for the unique needs of the project.

8.1.1.2 Description

Figure 8-1 is a notional graph for balancing formal process against the risk of cost and schedule overruns. As discussed in Chapter 2, insufficient SE effort is generally accompanied by high risk. Steven McConnell, in his book *The Power of Process,*[1] describes the improvements in efficiency realized by adding process. But as Figure 8-1 illustrates, too much formal process also introduces high risk. If too many or unnecessary processes are performed, increased cost and schedule impacts will occur with little or no added value to the integrity of the system.

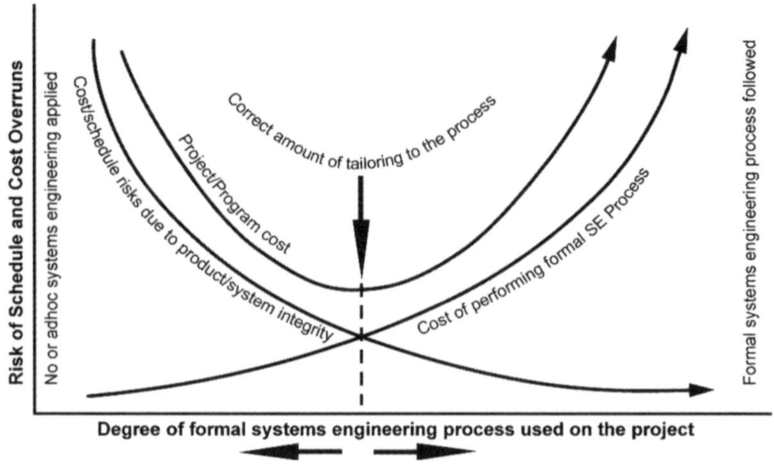

Figure 8-1 Tailoring requires balance between risk and process[2]

Tailoring organization and SE processes occurs dynamically over the system life cycle depending on risk and the situational environment and should be continually monitored and adjusted as needed. Figure 8-2 is the Context Diagram for the Tailoring Process described in this chapter.

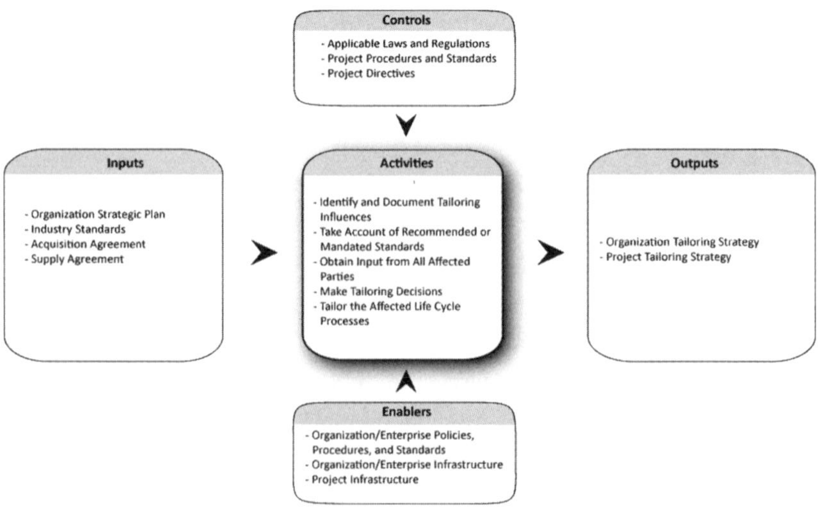

Figure 8-2 Tailoring Process Context Diagram

8.1.1.3 Inputs

Inputs to the Tailoring Process include the following:

- *Organization Strategic Plan* – Decision-making is governed by the organization strategic plan.

- *Industry Standards* – Relevant industry specifications and standards
- *Acquisition Agreement* – Terms and conditions of the agreements for acquired systems
- *Supply Agreement* – Terms and conditions of the agreement for the supplied system.

This process is governed by the following controls and enablers:

- *Applicable Laws and Regulations*
- *Project Procedures and Standards* – including project plans
- *Project Directives*
- *Organization/Enterprise Policies, Procedures, and Standards* – including guidelines and reporting mechanisms
- *Organization/Enterprise Infrastructure*
- *Project Infrastructure.*

8.1.1.4 Outputs

Outputs of the Tailoring Process include the following:

- *Organization Tailoring Strategy* – To the organization-level team responsible for incorporating and tailoring external standards into the organization's set of standard life cycle processes
- *Project Tailoring Strategy* – To the project-level team responsible for incorporating and tailoring the organization's set of standard life cycle processes for a given project.

8.1.1.5 Process Activities

The Tailoring Process includes the following activities:

- *Identify and Document Tailoring Influences*
  - Identify tailoring criteria for each stage – This activity establishes the criteria for including or excluding any process in the formal conduct of a given stage. Some essential processes, such as Configuration Management (see Section 5.5), build cumulatively throughout the system life cycle and may determine a set of permanent activities. Other processes, such as Project Planning (see Section 5.1), have a more limited range of applicability.
- *Take Account of Recommended or Mandated Standards*

- *Obtain Input from All Affected Parties*
  - *Determine process relevance to cost, schedule, and risks* – This activity analyzes the various environments, including their decision processes, relationships, and sensitivity to risks. The results define the appropriate tailoring of the review, decision, and coordination methods for each process activity in each stage.
  - *Determine process relevance to system integrity* – This activity analyzes the system features, intended environment, criticality of product/system use, reliability, and availability. It also defines the appropriate tailoring of the process activities, such as verification, qualification, level of analysis needed, and review and decision gate criteria.
  - *Determine quality of documentation needed* – This activity analyzes the support environment, system evolution, criticality of system functions, and internal and external interfaces. It defines the extent of detail needed in documentation for the project.
  - *Determine the extent of review, coordination and decision methods* – This activity analyzes project issues, such as stakeholder diversity, extent of their involvement, and nature of working relationships (e.g., single, unified, or conflicting customer needs). These factors influence tailoring of formal reviews, coordination and decision methods, and communications to fit the situation.
- *Make Tailoring Decisions*
- *Tailor the Affected Life Cycle Processes.*

Common approaches and tips:

- Tailor involves adding additional processes or eliminating unnecessary process. The following are examples of improper reasons for tailoring:
  - Not doing things because you do not want to do them
  - Not doing things because they are too hard
  - Not doing things because your boss doesn't like them.
- Tailoring decisions should be based on facts and should be approved by an independent authority.

- Tailoring Process activities should be conducted at least once for each stage of the system life cycle.
- Tailoring is driven by the environment of the system life-cycle stages.
- Agreements between organizations create constraints on tailoring.
- Issues of compliance to stakeholder, customer, and organization policies, objectives, and legal requirements will sometimes control the extent of tailoring.
- Certain documents and procedures may be mandatory in some situations.
- Methods of procurement or intellectual property will influence the extent of tailoring of the Agreement Process activities. Tolerance for formal processes is influenced by the contracting method – fixed price, cost plus fixed fee, time, and material.
- The life-cycle process/model used determines the extent and nature of SE process application, such as the number of reviews, development iterations, or decision points.
- Each participating organization brings their tolerance for risk to the Tailoring Process. Risk averse organizations may need more detailed information than what the system requires to build confidence in the processes. In such instances, tailoring may introduce extra activities that are removed as the level of trust builds between parties.
- A set of formal processes and activities are identified at the end of the Tailoring Process. This includes, but is not limited to:
  - A documented set of tailored processes
  - Identification of the system documentation required
  - Identified reviews
  - Decision methods and criteria
  - The analysis approach to be used.
- The tailoring plan, processes, documentation, and analyses are sensitive to change and increased knowledge from experience. By identifying the assumptions and criteria for tailoring, the Tailoring Process can be conducted throughout the life cycle to optimize the use of formal processes.

8.1.2 Elaboration

8.1.2.1 Organizational Tailoring

As indicated in Section 7.1, Life-Cycle Model Management provides the context for most processes by establishing standards, policies, processes, goals, and objectives that are based on market environment and opportunities, governmental and other external laws and regulations, and organization strategies in response to these factors. The organization context includes the industry domain. Influences on tailoring at the organizational level include:

- *Organization issues* – The organization environment controls the processes used in the development, determines who needs to approve certain products, defines what form and content the product takes, and what information can (or cannot) be shared between entities, both internal and external.
- *Organizational learning* – A key enabler in the Tailoring Process is experience with similar systems or familiarity between the participating parties. Beginning with less formal process structure for well-understood systems and established teams may yield significant cost savings without jeopardizing performance or quality.
- *Organizational maturity* – Established and well documented processes that are used frequently among parties can contribute to successful outcomes. In such instances, it may be more disruptive and add cost to remove such processes. Considering the maturity of the participating parties, both individually and as a whole, is an important enabler for tailoring.

When contemplating if and how to incorporate a new or updated external standard into and organization, the following should be considered:[3]

- Understand the Organization
- Understand the New Standard
- Adapt the Standard to the Organization (Not Vice Versa)
- Institutionalize Standards Compliance at the "Right" Level
- Allow for Tailoring.

8.1.2.2 Project Tailoring

Project tailoring applies specifically to the work executed through programs and projects. Factors that influence tailoring at the project level include:

- Stakeholders and customers (e.g., number of stakeholders, quality of working relationships, etc.)
- Project budget, schedule, and requirements
- Risk tolerance
- Complexity and precedence of the system.

Today's systems are more often an integration of many systems and system elements to create an operational environment. This demands that cooperation transcend the boundaries of any one organization. Harmony between multiple suppliers is often best maintained by agreeing to follow a set of consistent processes and standards. In such environments, consensus on a set of practices is helpful but adds complexity to the Tailoring Process.

8.1.2.3 Traps in Tailoring

Common traps in the Tailoring Process include, but are not limited to, the following:

1. *Reuse of a tailored baseline from another system without repeating the Tailoring Process* – It is fallacious to assume that previously tailored baselines are appropriate for all systems. Prior successes are not a guarantee of future success. There is something unique in each system.
2. *Using all processes and activities "just to be safe"* – The trap is that the each process carries an overhead cost. If this approach is taken, the quality of the system may actually degrade because of application of an inappropriate process. It cannot be called tailoring if there is not a clear justification for the inclusion (or exclusion) of every process in the plan.
3. *Using a pre-established tailored baseline* – Organization shortcuts to create baseline templates that can be taken off the shelf and applied to work based on arbitrary categorizations, such as high, medium, and low risk systems, can be counter-productive and carry the same hazards as traps #1 and #2. Tailoring is important because the emphasis is placed on the system, and only the processes that support attainment of the objective in terms of quality and performance should be retained.
4. *Failure to include relevant stakeholders* – The Tailoring Process itself can become a unifying activity that establishes shared visions and understanding of the objectives. Suppliers, or other organizations, that are identified and not included in the process may feel disenfranchised with the result that they feel a lower level of commitment to the process baseline. When new parties are added, they should be

familiarized with the baseline and asked to make constructive contributions.

8.2 References

1. McConnel, Steven, "The Power of Process," *IEEE Computer*, 1998, <http://www.stevemcconnell.com/articles/art09.htm>
2. Adapted from a presentation given by Ken Salter, at the Jet Propulsion Laboratory in Pasadena CA, 2003
3. Walden, D., "YADSES: Yet Another Darn Systems Engineering Standard," *Proceedings of the Seventeenth Annual International Symposium of the International Council on Systems Engineering*, San Diego, California, INCOSE, June 2007

9 Specialty Engineering Activities

The objective of this chapter is to give enough information to systems engineers to appreciate the significance of various engineering specialty areas, even if they are not an expert in the subject. It is recommended that subject matter experts are consulted and assigned as appropriate to conduct specialty engineering analysis. The topics in this chapter are covered in alphabetical order by topic title to avoid giving more weight to one topic over another. More information about each specialty area can be found in references to external sources.

With a few exceptions, the forms of analysis presented herein are similar to those associated with SE. Most analysis methods are based on the construction and exploration of models that address specialized engineering areas, such as electro-magnetic compatibility, reliability, safety, and security. Not every kind of analysis and associated model will be applicable to every application domain.

9.1 Design for Acquisition Logistics – Integrated Logistics Support[1,2]

The Operation and Maintenance Processes are defined in Sections 4.9 and 4.10, respectively. The sustainment of these processes during the Utilization and Support Stages is dependent on actions set in motion during the earlier stages. The sustainment of these processes and the influence to design from a whole life perspective is dependent on the application of a systematic set of activities and processes. These activities are known under various titles; Integrated Logistic Support (ILS) has a government/defense heritage while Supply Chain Management has a commercial heritage. Both have been typically established from a customer or product support background. The full scope of ILS includes Acquisition Logistics (i.e., technical and management activities conducted to ensure supportability implications are considered early and throughout the Acquisition Process) and Operational Logistics (i.e., activities that ensure that the right material and resources, in the right quantity and quality, are available at the right place and time throughout the Utilization and Support Stages). Operational logistics also receives attention under the heading of supply chain management. This handbook uses the term ILS, and this section focuses on Acquisition Logistics. Strategies and tactical plans for ILS are established as part of the Agreement Processes (see Chapter 6), and will drive the basic considerations to be applied during Acquisition Logistics activities.

Many different analyses are used to consider whether it is more cost effective to influence the initial design of the system or to plan for spare parts and repairs during utilization. When initial acquisition costs are fixed, this can have a downstream impact on the funding that will be needed in future years. These analyses also result in the determination of initial provisioning and spares.

9.1.1 "-ilities" Influencing the System Design

Acquisition Logistics focus on design requirements criteria applicable to all system elements. Sometimes referred to as the "-ilities," these criteria comprise, but are not limited to, the following list of engineering specializations: Affordability (LCC); Cost/System Effectiveness; Disposability (Recycling/Retirement); Maintainability; PHS&T; Producibility (Manufacturability); Reconfigurability (Flexibility/Standardization); Reliability; Security; Supportability (Serviceability); Survivability; and Vulnerability –" Figure 9-1 illustrates the relationship between ILS analysis activities.

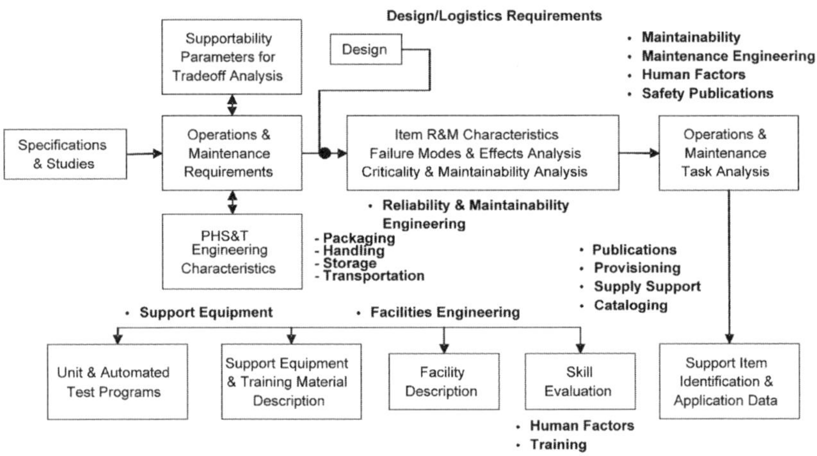

Figure 9-1 Acquisition Logistics Activities

The Reliability, Availability, and Maintainability of a system are major drivers in the use of support resources and the related in-service costs. Reliability is concerned with the probability of the system-of-interest working when it should. Maintainability is concerned with keeping the system working and the ease of putting things right once they have gone wrong.[3]

The probability that the system, when used under stated conditions, will operate satisfactorily is often expressed as Availability, which in turn, is dependent on the design for Reliability and Maintainability as well as the support arrangements during the Utilization and Support Stages. However, the term availability is a function of operating time (reliability) and downtime (maintainability/supportability).There are typically three categories of availability that are often expressed as requirements:

- Operational availability is the most difficult and includes logistics and administrative delay times that are not typically under the control of suppliers.
- Inherent availability is the simplest as it includes only the inherent R&M characteristics of the item under analysis.
- Measured availability is also complex as it requires the effective measurement of failures, their cause and the subsequent restorative action (maintenance). This also requires that effective Measurement Processes and systems are in place that are acceptable to the user, acquirer, and supplier stakeholders as part of an agreed feedback process.

A summarizing denomination for Reliability, Availability, Maintainability, and Supportability may also be known as Dependability. The Dependability of a system describes its ability to operate at certain points during a mission given conditions at the start of the mission, while taking degradation of performance due to failure and maintenance into consideration. Under certain conditions, it may be necessary to introduce redundancy into the design of a system to enhance the system's reliability by providing two or more functional paths or physical objects in areas that are critical to successfully achieve specified Availability under given conditions. Likewise, it is important to eliminate single points of failure during the design of a system.

In the discussion of the Øresund Bridge case (see Section 3.6.2), the survivability of the bridge in the event of ship collisions and analysis of the design to ensure that traffic flow is not interrupted by accidents are two examples of ILS-related trade studies that were conducted.

Another important factor to consider during the design of a system is PHS&T, which includes all special provisions, materials and containers and how the system or the parts thereof shall be handled, distributed, and stored. In addition to the system itself, PHS&T also covers spares and consumables.

- *Packaging* can occasionally be more expensive than the product itself, and packaging requirements often have conflicting user objectives. A merchant wants the package to be so tough that a shoplifter cannot steal parts from inside the package while it is on the store shelf. Conversely, the consumer wants the package to be easy to open when they get it home. Experience as a consumer suggests that the merchant won; many packages are almost impossible to open without using tools.
- *Handling* can also be a source of failure. A warehouse worker can – and sometimes does – drive a folk-lift through a package. Airline

baggage handlers are known to throw boxes onto carts, even though they are marked "fragile."

- *Storage* environments can impose significant constraints on packaging and the system itself. Humidity, dust, and temperature are typical environmental concerns. In addition, warehousing places constraints on size and density of objects.
- *Transportation* is often overlooked when designing a system. Use of freight trains or cargo planes imposes limits on height, width, length, and weight. For some systems, these constraints may force building multiple system elements that are not assembled until delivered to the operational site.

Acquisition Logistics/ILS is important and relevant to systems engineers. As consumers themselves, systems engineers have experienced the frustration of unreliable systems. Those who have maintained systems at some point in their careers know that failure to acknowledge that humans maintain systems can cause significant problems in an operational environment. When availability gets too bad, systems are sometimes simply shut down in frustration, which can cause significant rework at considerable cost. The huge variety of Acquisition Logistics analyses are best carried out by subject matter experts, because they are familiar with the mathematics that underlie the techniques, with the tools that are available to support these analyses, and with the factors that influence the outcome of these analyses.

9.1.2 "-ilities" Analysis Methods

The fields of Reliability, Maintainability and Supportability Engineering iteratively and recursively perform several analyses that are co-dependent on the results of other analyses. This section briefly addresses some of the most useful and common analyses techniques.

9.1.2.1 Failure Modes Effects and Criticality Analysis

Failure Modes Effects and Criticality Analysis (FMECA) is a means of recording and determining the following:

- What functions the equipment is required to perform
- How these functions could fail
- Possible causes of the failures
- Effects the failures would have on the equipment or system
- The criticality of the failures.

FMECA should be performed early enough to influence equipment design. The aim is to minimize maintenance requirements and thereby cost. FMECA indicates that potential failures may occur that either: cannot be removed through re-design but can be avoided through preventive maintenance; or have a non-critical impact and therefore can be allowed to occur, with subsequent rectification through corrective maintenance.

9.1.2.2 Level of Repair Analysis

Level of Repair Analysis is the process of evaluating system elements to first determine (in most cases, from an economic point of view) if the element or system should be discarded or repaired. If repairing the item is feasible, establish where the repair should take place (e.g., at home, locally, or at the factory), expressed as an organizational level. This analysis is conducted throughout the system life cycle. The handling of a system element may change based on experiences from prior decisions.

9.1.2.3 Logistic Support Analysis/Supportability Analysis

Logistic Support Analysis (LSA)/Supportability Analysis is a structured method of analyzing the support implications of system elements as they are being developed, with the aim of identifying features of the design that could result in excessive expense during the operational life of the system. Once identified, these items can be the subject of trade-offs to revise the design to reduce later costs. Once the design is more fully defined during the late activities of the Development Stage, the LSA can identify all the logistical resources necessary to support the equipment and the impact on the existing support infrastructure. LSA is only cost effective where it is likely to generate benefit in terms of a more supportable design or better defined support requirements and, hence, reduced LCC.

9.1.2.4 Reliability Centered Maintenance Analysis

Reliability Centered Maintenance (RCM) Analysis can be performed to assess the most cost efficient preventive maintenance program for the system. RCM is best initiated very early in the Development Stage and evolves throughout the Production Stage. As such, it can also successfully be introduced for systems already in operation, as it can be accomplished using a decision tree to lead the analyst through a logical sequence of the nature and frequency of applicable preventive maintenance tasks.

9.1.2.5 Survivability Analysis

Survivability Analysis is performed when items must perform critical functions in a hostile operational environment. Threats to be considered include conventional, electronic, nuclear, biological, chemical, and other weapons, as well as terrorism or sabotage, erratic human behavior, and harsh environmental conditions, such as ocean salinity. Critical survivability characteristics are identified, assessed, and analyzed to evaluate their impact on system performance and effectiveness.[4] A system is said to be survivable if it can fulfill its purpose in a timely manner, even in the presence of attacks or failures. Because of the severe consequences of failure, organizations increasingly focus on system survivability as a key risk topic.

The Spitfire (see Figure 9-2) was designed with an elliptical wing, giving greater speed and maneuverability (perhaps the most critical -ility of all for a warplane). But this came at a price: 13000 man-hours per airframe. Willy Messerschmitt had optimized the German Bf 109 for speed and manufacturability at only 4000 man-hours per frame, but the Bf 109 was no faster than the Spitfire and was consistently out-turned by it. The elliptical wing had been considered but ultimately rejected as too difficult to manufacture.[5]

Figure 9-2 The Spitfire: A perfect balance of -ilities?

9.1.2.6 System Security Analysis

System Security Analysis identifies and evaluates system vulnerabilities to known or postulated security threats and recommends means to eliminate the vulnerabilities or to at least reduce the susceptibility to compromise, damage, or destruction to an acceptable level of risk.

9.2 Cost-Effectiveness Analysis

In economics, the term cost-effectiveness applies to the comparison of the relative spending (costs) and outcomes (effects) associated with two or more courses of action. System cost-effectiveness analysis helps derive critical system performance and design requirements, and supports decision-making. Some examples of critical cost-effectiveness analyses are:

1. Studies of the desirable performance characteristics of commercial aircraft to increase an airline's market share at lowest overall cost over its route structure (e.g., more passengers, better fuel consumption)
2. Studies of the desired characteristics of a communications satellite to serve specified markets most economically (e.g., placement, coverage)

3. Urban studies of the most cost-effective improvements to a city's transportation infrastructure (e.g., buses, trains, motorways, and mass transit routes and departure schedules).

Military and government acquisitions are under the scrutiny of auditing offices to demonstrate that the money spent has delivered the expected benefits.[6] A recent concept, Cost-As-An-Independent-Variable builds on cost-effectiveness studies to determine an objective cost for the system acquisition. Once the cost is agreed upon, it becomes a constraint on future decisions regarding project execution.[7]

9.3 Electromagnetic Compatibility Analysis

Electromagnetic compatibility (EMC) analysis is performed on electric or electronic items to ensure that they can perform in their intended electromagnetic environments. Analysis also ensures that items that intentionally radiate radio frequency energy comply with commercial, government, and relevant international policies for radio frequency spectrum management and do not interfere with other signals (i.e., EMI). Even cable or speaker wire routing for home devices, such as a television, must consider EMC/EMI to achieve maximum performance and ensure safety of the users.

9.4 Environmental Impact Analysis

Europe, the United States, and many other nations recognize regulations that control and restrict the environmental impact that a system may inflict on the biosphere. The ISO 14000 series of Environmental Management[8] standards are an excellent resource for analysis and assessment methods for the protection of the environment. Failure to comply with environmental protection laws carries penalties and may result in the system not being approved for development. This issue is discussed in several references.[9,10]

The focus of environmental impact analysis is on potential deleterious effects of a proposed system's development, construction, use, and disposal. All countries that have legally expressed their concern for the environment restrict the use of hazardous materials (e.g., mercury, lead, cadmium, chromium 6, and radioactive materials) with a potential to cause human disease or to threaten endangered species through loss of habitat or impaired reproduction. Concern extends over the full life cycle of the system to be developed, as is made evident by the European Union's 2006 resolution to adopt a legal restriction that system developers and their suppliers retain lifetime liability for decommissioning systems that they build and sell.

The Øresund Bridge (see Section 3.6.2) is an example of how early analysis of potential environmental impacts ensures that measures are taken in the design and construction to protect the environment with positive results. Two key elements of the success of this initiative were the continual monitoring of the environmental status and the integration of environmental concerns into the requirements from the owner.

Disposal analysis is a significant analysis area within environmental impact analysis. Traditional landfills for non-hazardous solid wastes have become less available within large city areas, and disposal often involves transporting the refuse to distant landfills at considerable expense. The use of incineration for disposal is often vigorously opposed by local communities and citizen committees, and poses the problem of ash disposal since the ash from incinerators is sometimes classified as hazardous waste. Local communities and governments around the world have been formulating significant new policies to deal with the disposal of non-hazardous and hazardous wastes.

One goal of the architecture design is to maximize the economic value of the system-elements' residue and minimize the generation of waste materials destined for disposal. Because of the potential liability that accompanies the disposal of hazardous and radioactive materials, the use of these materials is carefully reviewed and alternatives used wherever and whenever possible. The basic tenet for dealing with hazardous waste is the "womb-to-tomb" control and responsibility for preventing unauthorized release of the material to the environment. This may include designing for reuse, recycling, or transformation (e.g., composing, bio-degradation).

In accordance with United States and European Union laws, system developers and supporting manufacturers must analyze the potential impacts of the systems that they construct and must submit the results of that analysis to government authorities for review and approval to build the system. Failure to conduct and submit the environmental impact analysis can result in severe penalties for the system developer and may result in an inability to build or deploy the system. It is best when performing environmental impact analysis to employ subject matter experts who are experienced in conducting such assessments and submitting them for government review.

9.5 Interoperability Analysis

Interoperability depends on the compatibility of elements of a large and complex system (which may sometimes be called a system-of-systems [SoS]; see Section 2.5] or a family of systems) to work as a single entity. This feature is increasingly important as the size and complexity of systems continues to grow. Pushed by an inexorable trend toward electronic digital systems and pulled by the accelerating pace of digital technology invention, commercial firms and national organizations span the world in increasing numbers. As their spans increases, these commercial and national organizations want to ensure that their sunken investment in legacy elements of the envisioned new system is protected and that new elements added over time will work seamlessly with the legacy elements to comprise a unified system.

Standards have also grown in number and complexity over time, yet compliance with standards remains one of the keys to interoperability. The standards that correspond to the layers of the *ISO-OSI Reference Model* for peer-to-peer communication systems once fit on a single wall chart of modest size. Today, it is no longer feasible to identify the number of standards that apply to the global communications network on a wall chart of any size. Interoperability will increase in importance as the world grows smaller due to expanding communications networks and as nations continue to perceive the need to communicate seamlessly across international coalitions of commercial organizations or national defense forces.

The Øresund Bridge (see Section 3.6.2) demonstrates the interoperability challenges faced when just two nations collaborate on a project, for example, the meshing of regulations on health and safety and the resolution of two power supply systems for the railway.

9.6 Life-Cycle Cost Analysis

As discussed in Chapter 2, decisions made during the early stages of a project inevitably have an impact on future expenditures. New systems are designed, developed, manufactured, and verified over the span of many years, as in the case of a new automobile, or nearly two decades, as in the case of a submarine. Over such lengths of time, decisions made at the outset may have substantial, long-term effects that are frequently difficult to analyze.

Life-cycle cost (LCC) analysis is a method of economic evaluation that takes into account all relevant costs of a system over a given period of time adjusting for differences in the timing of those costs.[11] For products purchased off the shelf, the major factors are the cost of acquisition, operation, maintenance, and disposal. Otherwise, it may be necessary to include the costs associated with each of the six life-cycle stages. An LCC analysis results in a timetable of expenses so that an organization can cover its costs. If all costs cannot be covered, it may not be possible to produce the system. The following activities are recommended for performing LCC analyses:

1. Obtain a complete definition of the system, elements, and their subsystems.
2. Determine the total number of units of each element, including operational units, prototypes, spares, and test units to be procured. If it is desired to develop parametric cost data as a function of the number of operational units, define the minimum and maximum

number of operational units and how, if any, the number of spares and test units will vary with operational unit size.

3. Obtain the life-cycle program schedule, including spans for R&D, Production and Deployment, and O&S stages. The Production and Deployment Stage length will vary with the number of operational units.

4. Obtain manpower estimates for each stage of the entire program and, if possible, for each element and subsystem. These are especially important for cost estimating during R&D and O&S.

5. Obtain approximate/actual overhead, general and administrative (G&A) burden rates, and fees that should be applied to hardware and manpower estimates. Usually, this is only necessary for effort within your own company; suppliers will have already factored it into their cost estimates. These data are not required to the accuracy that your finance department would use in preparing a formal cost proposal.

6. Develop cost estimates for each subsystem of each system element for each stage of the program. This is, of course, the critical step. Generally, it should be done as accurately as time and resources allow. Common methods/techniques for conducting LCC analyses are as follows:

    a. *Expert Judgment* – Consultation with one or more experts. Good for sanity check, but may not be consistent.

    b. *Analogy* – Reasoning by comparing the proposed project with one or more completed projects that are judged to be similar, with corrections added for known differences. May be acceptable for early estimations.

    c. *Parkinson Technique* – Defines work to fit the available resources.

    d. *Price-To-Win* – Focuses on providing an estimate, and associated solution, at or below the price judged necessary to win the contract.

    e. *Top-Down* – Based on developing costs from the overall characteristics of the project from the top level of the architecture.

    f. *Bottom-Up* – Identifies and estimates costs for each element separately and sums the contributions.

g. *Algorithmic (parametric)* – Uses mathematical algorithms to produce cost-estimates as a function of cost-driver variables, based on historical data. This technique is supported by commercial tools and models.

h. *Design-to-Cost or Cost-As-An-Independent-Variable* – Works on a design solution that stays within a predetermined set of resources.

i. *Wide-band Delphi techniques* – Builds estimates from multiple technical and domain experts. Estimates are only as good as the experts.

j. *Taxonomy method* – Hierarchical structure or classification scheme for the architecture.

Sometimes the argument is heard that the LCC estimates are only to support internal program tradeoff decisions and, therefore, must only be accurate enough to support the tradeoffs (relative accuracy) and not necessarily realistic (absolute accuracy). This is usually a bad practice and, if done, is a risk element that should be tracked for some resolution of veracity. The analyst should always attempt to prepare accurate cost estimates. These estimates are often reviewed by upper management and customers. The credibility of results is significantly enhanced if reviewers sense the costs are "about right," based on their past experience.

7. Document the results of LCC analysts. Adequate documentation requires three basic elements:

   a. Data and sources of data on which the estimate is based

   b. Estimating methods applied to that data

   c. Results of the analysis.

Life-cycle cost analyses are used in system cost-effectiveness assessments, as discussed in Section 9.2. The LCC is <u>not</u> necessarily the definitive cost proposal for a program Since LCC estimates are often prepared early in a program's life cycle—during Concept Definition. At this stage, there is insufficient detailed design information available to support preparation of a realistic, definitive cost analysis. These analyses are much more detailed and prepared perhaps several years later than the earliest LCC estimates. Later in the program, LCC estimates can be updated with actual costs from early program stages and should be more definitive and accurate due to hands-on experience with the system. In addition to providing information for the LCC estimate, these studies also help to identify areas in which emphasis can be placed during the subsequent substages to obtain the maximum cost reduction.

In some literature, LCC is equated to total-cost-of-ownership. LCC analysis helps the project manager understand the total cost impact of a decision, compare between program alternatives, and support trade studies for decisions made *throughout* the system life cycle. LCC normally includes the following costs, represented in Figure 9-3.

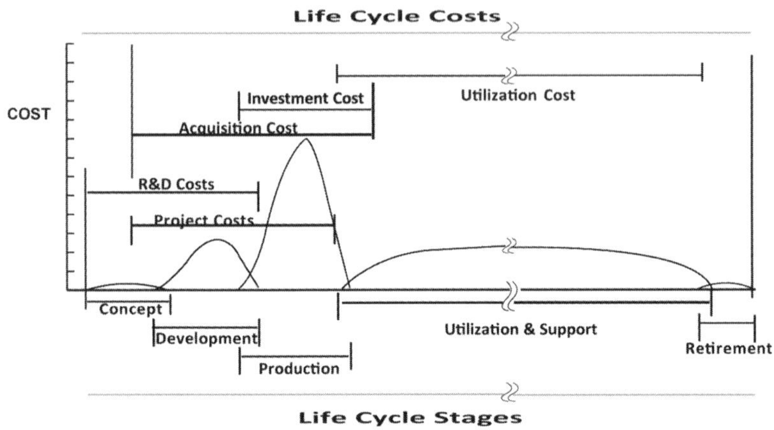

Figure 9-3 Life-Cycle Cost Elements (not to scale)

Accuracy in the estimates will improve as the system evolves and the data used in the calculation is less uncertain.

1. *R&D and O&S costs* – Can usually be estimated based on average manpower and schedule spans and include overhead, G&A costs, and fees, as necessary.
2. *Investment costs* – Usually prepared by estimating the cost of the first production unit, then applying learning curve formulae to determine the reduced costs of subsequent production units. For an item produced with a 90 percent learning curve, each time the production lot size doubles (2, 4, 8, 16, 32, ... etc.) the average cost of units in the lot is 90 percent of the average costs of units in the previous lot. A production cost specialist is usually required to estimate the appropriate learning curve factor(s).
3. *Utilization and Support costs*
4. *Disposal costs.*

R&D and Investment costs can sometimes be scaled by "complexity factors" or Cost Estimating Relationships (CERs) from accurate costs of existing items. This entails fact finding with experts familiar with the item.

A typical example of a decision with a long-term effect concerns the choice of system elements and parts for a new system. Often, the desire to minimize the initial investment by selecting less expensive parts carries the consequence of a higher probability of failure during the operational life of the system and the corresponding higher maintenance costs. Therac-25 (see Section 3.6.1) is an example where the decision to save time and money on software unit verification resulted in undetected design errors that emerged later during operation.

Another example relates to the need for logistics support. Maintenance costs during the Operational Stage can be reduced by including in the system built-in test equipment that identifies problems, locates their source, and recommends a corrective course of action. Diagnostic testing elements of this type that combine sensors with automated checklists and expert systems logic are expensive to develop, but in the long run decrease maintenance costs and increase availability. A relatively small change in mean-time-to-repair or mean-time-between-failures can result in large cost savings during the Operations Stage.

LCC analysis has three important benefits:

- All costs associated with a system become visible: upstream; locked in costs, such as R&D; downstream; customer service.[12]
- Supports an analysis of organization interrelationships. Reinforces the importance of locked in costs, such as R&D; low R&D expenditures may lead to high customer service costs in the future.
- Project managers can develop accurate revenue predictions.

9.7 Manufacturing and Producibility Analysis

The capability to produce a system element is as essential as the ability to properly define and design it. A designed product that cannot be manufactured causes design rework and program delays with concomitant cost overruns. For this reason, production engineering analysis and trade studies for each design alternative form an integral part of the Architectural Design Process. One objective is to determine if existing proven processes are satisfactory since this could be the lowest risk and most cost-effective approach. The Maglev train contractor (see Section 3.6.3) experienced a steep learning curve to produce an unprecedented system from scientific theory.

Producibility analysis is a key task in developing low-cost, quality products. Multidisciplinary teams work to simplify the design and stabilize the manufacturing process to reduce risk, manufacturing cost, lead time, and cycle time and to minimize strategic or critical materials use. Critical producibility requirements are identified during system analysis and design and included in the program risk analysis, if necessary. Similarly, long-lead-time items, material limitations, special processes, and manufacturing constraints are evaluated. Design simplification also considers ready assembly and disassembly for ease of maintenance and preservation of material for recycling. When production engineering requirements create a constraint on the design, they are communicated and documented. The selection of manufacturing methods and processes are included in early decisions.

Manufacturing analyses draw upon the Concept of Production, Concept of Deployment, and Concept of Maintenance. Manufacturing test considerations are shared with the engineering team and are taken into account in Built-In-Test and Automated Test Equipment.

IKEA® is often used as an example of supply chain excellence. IKEA® has orchestrated a value creating chain that begins with motivating customers to perform the final stages of furniture assembly in exchange for lower prices and a fun shopping experience. They achieve this through designs that support low cost production and transportability (e.g., the bookcase that comes in a flat package and goes home on the roof of a car).

9.8 Mass Properties Engineering Analysis

Mass Properties Engineering[13] (MPE) ensures that the system or system element has the appropriate mass properties to meet the requirements. Mass properties include weight, the location of center of gravity, inertia about the center of gravity, and product of the inertia about an axis.

Typically, the initial sizing of the physical system is derived from other requirements, such as minimum payload, maximum operating weight, or human factors restrictions. Mass properties estimates are made at all stages of the system life cycle based on the information that is available at the time. This information may range from parametric equations to a three-dimensional product model to actual inventories of the product in service. A risk assessment is conducted, using techniques such as uncertainty analysis or Monte Carlo simulations, to verify that the predicted mass properties of the system will meet the requirements and that the system will operate within its design limits. Validation (see Section 4.8) is usually conducted at the end of the Production Stage to assure all parties that the delivered system meets the requirements, and then several times during the Utilization Stage to ensure the safety of the system, system element, or human operator. For a multi-billion dollar project, such as oil platform or warship, the MPE level of effort is significant.

One trap in MPE is that design managers may believe that their 3-D modeling tools can be used to estimate the mass properties of the system or system element. This is problematic because (1) not all parts are modeled on the same schedule and (2) most parts are modeled neat, that is, without such items as manufacturing tolerances, paint, insulation, fittings, etc., which can add from 10 to 100% to the system weight. For example, the liquid in piping and tanks can weigh more than the structural tank or metallic piping that contain it.

Mass Properties Engineering usually includes a reasonableness check of all estimates using an alternative method. The simplest method is to justify the change between the current estimate and any prior estimates for the same system or the same system element on another project. Another approach is to use a simpler estimating method to repeat the estimate then justify any difference.

9.9 Safety & Health Hazard Analysis

Safety and health hazards are hazards to the well-being of human operators, maintainers, administrators, or other system users. They are a major concern[14] wherever hazardous materials are employed, such as the chemical industries, building organizations, medical and radiological equipment supply industry, energy production, and aviation and space. The objective of the system safety effort is to ensure that the system meets an accepted level of risk. Assessing safety risk is almost always necessary for contractual and legal reasons.

A systems engineer in one of the cited industries or any number of other industries that deal with hazardous materials, processes, or human activities must be aware that SMEs and system safety engineers are available to perform the analyses that can identify these hazards and their attendant risks. As such, the SE and SMEs can help identify means to eliminate or at least mitigate safety and health risks to acceptable levels. The system safety program begins in the conceptual stage, and continues throughout the system life cycle, as illustrated in Figure 9-4.

Concept Stage		Development Stage		Production Stage	Utilization Stage / Support Stage		Retirement Stage
Establish Safety Objectives	Initial Hazard Analysis	Update Hazard Analysis	Safety Verification	Sys Safety Assessment & Certification	Maintenance of Safety Baseline	Mishap Investigation and Correction	Safe Disposal

Figure 9-4 System safety focus during the system life cycle

Safety risks are associated with such processes as complex machinery used in a manufacturing plant, high-temperature metals in a steel plant, coal mining, or maintenance of deep sea platforms; or with activities such as flying, space travel, or deep sea fishing. While a safety decision tree can be a useful starting place to analyze processes and activities as well as physical elements of systems, it is likely that the means to eliminate or reduce process and activity risks will be different. Construction of safety cages can protect people in a complex manufacturing work cell; "kill" buttons can be installed; and barriers can be constructed to make sure a person cannot fall (for example) into molten steel. Specialized training and back-up safety equipment can also be made available to, for example, divers that maintain off-shore oil rigs. The Therac-25 case (see Section 3.6.1) illustrates the cost in human life that may result when adequate measures are not taken to build safety measures into potentially dangerous equipment. Conversely, the specially designed windows for the Maglev train (see Section 3.6.3) dampen the noise level that would otherwise present a hazard to passengers.

When the hazards are caused by materials used within the system, it is crucially important to isolate the materials by some safe means as they are used in the system, and to plan for their eventual substitution by non-hazardous materials as material science advances. Figure 9-5 shows examples of protective clothing used in the radiological, medical, and other hazardous materials industries.

Figure 9-5 Protective clothing for Hazmat Level A and bird flu

Many governments have regulations that mandate that all hazards to human safety and health be reduced as far as is possible, and that all safety and health hazards that cannot be eliminated are mitigated by other than system means (e.g., protective clothing, etc.) to reach acceptable levels of risk. This means avoiding wherever possible the use of hazardous materials, containing hazardous material that cannot be eliminated, and addressing the hazards associated with process and human activities that are required to support and maintain the system in its operational environment. It also means planning for the safe handling and disposal of hazardous materials, and including such effort in the LCC models and cost forecasts for the system being developed.

9.10 Sustainment Engineering Analysis

Sustainment engineering helps ensure that a system continues to satisfy its objective over its intended lifetime. In that timeframe, system expectations will expand, environments in which the system is operated will change, technology will evolve, and elements of the system may become unsupportable and need to be replaced. The desktop computing environment is a case in point. Since the introduction of the Universal Serial Bus (USB), it is nearly impossible to find cables to support parallel port printers.

Sustainment Engineering is an integrated effort designed to address industry needs regarding aging systems and a need to maintain those systems in operation. A sustainment program may include re-engineering electronic and mechanical elements to cope with obsolescence, the development of automated test equipment, and extending the life of aging systems through technology insertion enhancements and proactive maintenance. These changes will have significant impact on ILS analyses (see Section 9.1).

9.11 Training Needs Analysis

Training needs analyses support the development of products and processes for training the users, maintainers, and support personnel of a system. Training analysis includes the development of personnel capabilities and proficiencies to accomplish tasks at any point in the system life cycle to the level they are tasked. These analyses address initial and follow-on training necessary to execute required tasks associated with system use and maintenance. An effective training analysis begins with a thorough understanding of the concept documents and the requirements for the system-of-interest. A specific list of functions or tasks can be identified from these sources and represented as learning objectives for operators, maintainers, administrators, and other system users. The learning objectives then determine the design and development of the training modules and their means of delivery.

Important considerations in the design of training include who, what, under what conditions, and how well each user must be trained, and what training will meet the objectives. Each of the required skills identified must be transformed into a positive learning experience and mapped onto an appropriate delivery mechanism. The formal classroom environment is rapidly being replaced with or augmented by simulators, computer-based-training, internet-based distance delivery, and in-systems electronic support, to name a few. Updates to training content use feedback from trainees after they have some experience to improve training effectiveness.

9.12 Usability Analysis/Human Systems Integration

Human Systems Integration is the interdisciplinary technical and management processes for integrating human considerations within and across all system elements. HSI focuses on the human, an integral element of every system, over the system life cycle. It is an essential enabler to SE practice as it promotes a "total system" approach that includes humans, technology (i.e., hardware, software), the operational context, and the necessary interfaces between and among the elements to make them all work in harmony[15,16,17,18,19] The "human" in HSI includes all personnel who interact with the system in any capacity:

- System owners
- Users/customers
- Operators
- Decision-makers
- Maintainers
- Support personnel
- Trainers
- etc.

Human Systems Integration brings human-centered disciplines and concerns into the SE process to improve the overall system design and performance. Thus, it is clear that the human is an element of every system, so all systems benefit from HSI application.

The primary objective of HSI is to ensure that human capabilities and limitations are treated as a critical system element, regardless of whether humans in the system operate as individuals, crews, teams, units, or organizations. The technology elements of the system have inherent capabilities. Similarly, humans possess particular knowledge, skills, abilities, expertise, and cultural experiences. HSI seeks to treat humans as essential system elements in design, just like other system elements (e.g., hardware and software). Deliberate design effort is essential to ensure development of quality interfaces between technology elements and the system's intended users, operators, maintainers, and support personnel in operational environments. It is also important to acknowledge that humans outside the system may be affected by its operation. Accordingly, any prescription for SE must accommodate these people as well.

While many systems engineers and design engineers intuitively understand that the human operator and maintainer are part of the system under development, they often lack the expertise or information needed to fully specify and incorporate human capabilities with the capabilities of the hardware and software. HSI assists systems engineers by bringing the various human-centered domains into the SE process and serving as the focal point to ensure that human considerations are integrated into system design, development, manufacturing, operation, sustainment, and disposal. This broad and disciplined approach focuses on important customer/user issues:

- Usability
- Usefulness
- Suitability
- Effectiveness
- Safety and health
- Resilience
- Understanding of the technological elements
- Reliability
- Availability
- Maintainability
- Supportability
- Trainability
- Cost of ownership.

The comprehensive application of HSI to system development, design, and acquisition is intended to optimize total system performance (i.e., human, hardware, and software), while accommodating the characteristics of the population that will use, operate, maintain, and support the system; and also support efforts to reduce total ownership costs.

Key methods of HSI are trade studies and analyses that address incorporation of human issues. IPDT analyses, especially requirements analyses, in which human issues and implications are explored with the users and other engineering disciplines, often result in insights not otherwise realized. Therefore, direct participation by HSI personnel in the engineering process is extremely valuable. Additionally, trade studies that include human-related issues are critical to optimize total system performance and determine the design that is the most effective, efficient, suitable (including useful and understandable), usable, safe, and affordable.

Systems development organizations routinely focus on short term acquisition cost and schedule, while not paying sufficient attention to the more expensive total ownership costs, for example:

- Personnel costs

- Repair and sustainment costs
- Cost of redesign and retrofit to correct deficiencies
- Training costs
- Cost of mishaps
- Handling hazardous materials
- Disability compensation and liability claims
- Environmental clean-up and disposal costs.

Long-term system success and customer satisfaction rely heavily upon demonstrated effectiveness of the total system inclusive of its decision-makers, operators, maintainers, supported customers, sustainers, and the support network. In all systems, failure to address long-term, life-cycle issues can result in failure to achieve the intended purpose/mission, a poor design, unnecessary burdens on the workers, increased incidence of human errors, excessive total cost of ownership, and, in some cases, negative impacts to the environment and public health and safety. Additionally, economic consequences may include lost customer confidence, lost market share, product liability, and little repeat business. Without this total system approach, the system as an enterprise solution will not meet optimal total system performance and/or total cost of ownership objectives. HSI helps systems engineers to focus on long-term costs since much of the total cost of ownership is related directly to human performance and the component areas of HSI.

An unforgettable example is the 1979 nuclear incident at the Three Mile Island power plant in the United States:

> *"The accident was caused by a combination of personnel error, design deficiencies, and component failures. The problems identified from careful analysis of the events during those days have led to permanent and sweeping changes in how NRC regulates its licensees – which, in turn, has reduced the risk to public health and safety." Further the article states: "it brought about sweeping changes involving emergency response planning, reactor operator training, human factors engineering, radiation protection, and many other areas of nuclear power plant operations."*[20]

In this example, the failure to do HSI within a comprehensive SE framework resulted in loss of confidence in nuclear power in the United States and delayed progress in the field for almost 30 years. The clean-up costs, legal liability, and significant resources associated with responding to this near catastrophe trace directly to lack of attention to the human element of a highly complex system, resulting in flawed operations technology and work methods. Three Mile Island also emphasizes that, while human performance includes raw efficiency in terms of tasks accomplished per unit time and accuracy, human performance also involves additional considerations, such as intrinsic motivation, the opportunity to learn and improve, and gaining confidence that the human-machine entity can cope with rare, surprising, or non-routine events. These considerations can be positively or negatively impacted depending upon how well the hardware and software support human capabilities and limitations in meeting system goals.

9.12.1 HSI is Integral to the SE Process

The foundation for program success is rooted in requirements development. Human performance requirements are derived from and bounded by other performance requirements within the system. Front-end analyses are extremely important in generating system requirements, which incorporate HSI-related requirements. Effective front-end analyses start with a thorough understanding of the mission of the new system and the actual work to be performed; successes or problems with any predecessor systems, systems with which the proposed system must interface; and the knowledge, skills, abilities, and training associated with the people who are likely to interact with the proposed system technology. Validated HSI modeling and simulation also can pay large dividends early in the development process, particularly before system hardware and software elements are developed and begin development/certification testing. Supported by this background information, HSI analyses are used to allocate human-centered functions within the system and identify any human capability gaps with recommended solutions. Decisions about whether or not to automate certain functions can be evaluated with modeling and simulation to identify and reduce risk, or at least scope the types and levels of risk involved.

HSI activities must begin with initial capabilities studies followed by continuing analyses (e.g., requirements and capability gap analyses, analyses of alternative solutions, task/function analyses, etc.). With a clearer understanding of the missions, functions, operational scenarios, and tasks that must be supported by the human and technological elements of the system, the SE team can help refine the overall system requirements, inclusive of HSI considerations.

It is critical to include HSI early in system development (during capabilities requirements generation) and continuously through the development process to realize the greatest benefit to the final system solution and substantial life cycle cost (LCC) savings. Changes made early in the process, when concepts and requirements are formulated, can have major positive impacts compared to later stages of system development where these same changes would be much more difficult and costly to implement with actual hardware and software.

Fully utilizing HSI in IPPD helps ensure that systems will not require expensive "train-arounds" or late-stage fixes to address issues of ineffective usability and inefficiencies due to poor human interfaces. The systems engineer plays a critical role in engaging HSI expertise to support the IPDTs, ensuring that human performance and interface considerations are properly identified and developed in the design and that sound functional allocations are made. A knowledgeable, interdisciplinary HSI team is generally required to address the full spectrum of human considerations, and the systems engineer is key to ensuring that HSI is included throughout the system's life cycle.

The attributes of communication, coordination, and collaboration are also essential to success of the development team. Program managers, chief engineers and systems engineers should ensure that HSI practitioners are actively participating in design reviews, working groups, and IPDTs. Consistent involvement and communications with customers, users, developers, scientists, testers, logisticians, engineers, and designers (human, hardware, and software) are essential to successful HSI and good SE practice. Integration takes serious effort, dedicated team participation, and willingness to make effective compromises to benefit the total system.

9.12.2 Technical and Management HSI Processes

Human Systems Integration must be addressed by all program level IPDTs and at program, technical, design, and decision reviews throughout the life of the system. HSI influences the design and acquisition of developmental systems, non-developmental systems, and system modifications, to include associated support requirements. HSI implementation makes explicit the role that the human plays in system performance and cost, and how these factors are shaped by design decisions. In addition, HSI is one of the essential components of engineering practice for system development, contributing technical and management support to the development process itself.

9.12.2.1 HSI Domains

The HSI processes facilitate trade-offs among the individual, but very interdependent, human-centric domains without replacing individual domain activities, responsibilities, or reporting channels. The human-centered domains typically cited by various organizations may differ in what they are called or in number, but the human considerations addressed are quite similar. The following human-centered domains with recognized application to HSI serve as a good foundation of human considerations that need to be addressed in system design and development, but clearly are not all inclusive:

- *Manpower* – Addresses the number and type of personnel in the various occupational specialties required and potentially available to train, operate, maintain, and support the deployed system. The manpower community promotes pursuit of engineering designs that optimize the efficient and economic use of manpower, keeping human resource costs at affordable levels. Determination of required manpower positions must recognize the evolving demands on humans (e.g., cognitive, physical, and physiological) and consider the impacts that technology can make on humans integrated into a system.

- *Personnel* – Considers the type of human knowledge, skills, abilities, experience levels, and human aptitudes (i.e., cognitive, physical, and sensory capabilities) required to operate, maintain, and support a system and the means to provide (i.e., recruit and retain) such people. Personnel recruitment, testing, qualification, and selection are driven by system requirements. The personnel community helps define the human performance characteristics of the user population and then determines target populations to select for occupational specialties, manage recruitment, and track retention trends. The personnel community manages occupational specialties to include career progression and assignments. Adequate numbers of workers in these specialties must be recruited, trained, and assigned to meet the entire career field need. Personnel population characteristics can impact manpower and training, as well as drive design requirements.
- *Training* – Encompasses the instruction and resources required to provide personnel with requisite knowledge, skills, and abilities to properly operate, maintain, and support systems. The Training community develops and delivers individual and collective qualification training programs, placing emphasis on options that:
  - Enhance user capabilities to include operator, maintainer, and support personnel)
  - Maintain skill proficiencies through continuation training and retraining
  - Expedite skill and knowledge attainment
  - Optimize the use of training resources.

  Training systems, such as simulators and trainers, should be developed in conjunction with the emerging system technology. The overall training system architecture is established from the Manpower and Personnel analyses, and the training delivery system may be required prior to fielding the system so that personnel can be adequately trained to operate, maintain, and support the system when it is fielded. Therefore, it also is important to develop the training system concurrent with the operational system. If engineering changes are made to the operational system, associated training architecture and delivery system changes must be re-evaluated, re-planned, and appropriate modifications funded.

- *Human Factors Engineering (HFE)* – Involves an understanding of human capabilities (i.e., cognitive, physical, sensory, and team dynamic) and comprehensive integration of those capabilities into system design beginning with conceptualization and continuing through system disposal. A key objective for HFE is to clearly characterize the actual work to be performed, and then use this information in creating effective, efficient, and safe human-hardware/software interfaces to achieve optimal total system performance (i.e., use, operation, maintenance, support, and sustainment). This "optimal performance" is the achievement of the following:
  - Conducting task analyses and design trade-off studies to optimize human activities creating work flow
  - Making the system intuitive to humans who will use, operate, maintain, and support it
  - Providing deliberately designed primary, secondary, backup, and emergency tasks and functions
  - Meeting or exceeding performance goals and objectives established for the system
  - Conducting analyses to eliminate/minimize the performance and safety risks leading to task errors and system mishaps across all expected operational, support, and maintenance environments.

Human Factors Engineering, through task and function analyses (including cognitive) task analysis, defines system functions and interfaces. These efforts should recognize the increasing complexity of technology and the associated demands on people, giving careful consideration to the capabilities and limitations of humans. The design should not demand unavailable or unachievable skills. HFE seeks to maximize usability for the targeted range of users/customers and to minimize design characteristics that induce frequent or critical errors. HFE works with the IPDTs to ensure that representative personnel are tested in situations to determine whether the human can operate, maintain, and support the system in adverse environments while working under the full range of anticipated mission stress and endurance conditions.

- *Environment* – In the context of HSI, this domain involves environmental considerations that can affect the ConOps and requirements, particularly human performance. These considerations include measures to protect the human element of the total system from the operational environment (i.e., shock, vibration, extreme temperatures, etc.) and to protect the environment (i.e., water, air, land, space, cyberspace, markets, organizations, and the relationships that exist among them and with all living things) from adverse effects associated with system development, manufacturing, operations, sustainment, and disposal activities.

- *Safety* – Promotes system design characteristics and procedures to minimize the risk of accidents or mishaps that cause death or injury to operators, maintainers, and support personnel; threaten the operation of the system; or cause cascading failures in other systems. Using safety analyses and lessons learned from predecessor systems, the Safety community prompts design features to prevent safety hazards to the greatest extent possible and to manage safety hazards that cannot be avoided. The focus is on designs that have back-up systems and, where an interface with humans exists, to alert them when problems arise and help the total system avoid and recover from errors. Prevalent issues include the following:

  - Factors that threaten the safety of personnel and their operation of the system
  - Walking/working surfaces, emergency egress pathways; personnel protection devices
  - Pressure and temperature extremes
  - Prevention/control of hazardous energy releases (e.g., mechanical, electrical, fluids under pressure, ionizing or non-ionizing radiation, fire, and explosions).

- *Occupational Health* – Promotes system design features and procedures that serve to minimize the risk of injury, acute or chronic illness, and disability and to enhance job performance of personnel who operate, maintain, or support the system. The Occupational Health community prompts design features to prevent health hazards where possible and recommends personal protective equipment, protective enclosures, or mitigation measures where health hazards cannot be avoided. Prevalent issues include the following:

  - Noise and hearing protection
  - Chemical exposures and skin protection

- Atmospheric hazards (e.g., Confined space entry and oxygen deficiency)
- Vibration, shock, acceleration, and motion protection
- Ionizing/non-ionizing radiation and personnel protection
- Human factors considerations that can result in chronic disease or discomfort (e.g., repetitive motion injuries or other ergonomic-related problems).

- *Habitability* – Involves characteristics of system living and working conditions, such as the following:
  - Lighting
  - Ventilation
  - Adequate space
  - Vibration, noise, and temperature control
  - Availability of medical care, food, and/or drink services
  - Suitable sleeping quarters, sanitation and personal hygiene facilities, and fitness/recreation facilities.

  Attention to such characteristics is necessary to sustain high levels of personnel morale, motivation, quality of life, safety, health, and comfort, contributing directly to personnel effectiveness and overall system performance. These habitability characteristics also directly impact personnel recruitment and retention. Some operational/organizational, technical, or mission issues may preclude completely addressing all habitability concerns: hence, other HSI domains may need to engage to mitigate the resulting effects on system personnel and performance.

- *Survivability* – Addresses human-related characteristics of a system (e.g., life support, body armor, helmets, plating, egress/ejection equipment, air bags, seat belts, electronic shielding, alarms, etc.) that reduce susceptibility of the total system to mission degradation or termination; injury or loss of life; and partial or complete loss of the system or any of its elements. These issues must be considered in the context of the full spectrum of anticipated operations and operational environments and for all people who will interact with the system (e.g., users/customers, operators, maintainers, or other support personnel). Adequate protection and escape systems must provide for personnel and system survivability when they are threatened with harm.

These individual, but very interdependent, domains must be considered simultaneously because decisions made in one domain can have significant impacts on other domains. Each individual domain decision generates the need to concurrently assess HSI issues across all the domains and against mission performance, prior to making formal programmatic decisions. This approach mitigates the potential for unintended, adverse consequences, including increased technical risk and cost.

9.12.2.2 Key HSI Activities and Tenets

Human Systems Integration programs have distilled the following HSI activities and associated key actionable tenets:

1. *Initiate HSI Early and Effectively* – HSI should begin in early system concept development with front-end analyses and requirements definition.
    - HSI-related requirements include not just those of the individual domains, but also those that arise from interactions among the HSI domains.
    - HSI requirements must be developed in consonance with other system requirements and consider any constraints or capability gaps. The human considerations identified in the requirements must address the capabilities and limitations of users, operators, maintainers, and other personnel as they interact with and within the system. HSI requirements must be reconsidered, refined, and revised as program documents, system requirements, and specifications are updated. Doing this early, continuously, and comprehensively as part of the SE process provides the opportunity to identify risks and costs associated with program decisions.
    - HSI must be included in the initial Acquisition Strategy, SEP, and LCC and life-cycle sustainment documents.
2. *Identify Issues and Plan Analysis* – Projects/programs must indentify HSI-related issues that require analysis to ensure thorough consideration.
    - The project/program manager must have a comprehensive plan for HSI early in the Acquisition Process and summarize HSI planning in the Acquisition Strategy. Project/program managers should address HSI throughout the entire acquisition cycle as part of the SE process.

- Efficient, timely, and effective planning/re-planning and front-end analyses are the cornerstones of HSI efforts, ensuring human considerations are effectively integrated into capability requirements, system concept development, and acquisition. HSI front-end analyses establish and assess criteria for success and help determine when a system design change is required.
- HSI planning should be integrated within the program's SEP or comparable planning document.
- HSI planning information should include the details associated with the analyses in SE language to support HSI tradeoffs, such as goals of the analyses and required timeline and resources.

3. *Document HSI Requirements* – Systems engineers derive HSI requirements, as needed at each level of the system hierarchy, using HSI plans, analyses, and reports as sources (or rationale) for the derived requirements.
    - Requirements tools and processes facilitate flow-down of high level HSI requirements to system elements.
    - HSI requirements should be cross-referenced with other documents, plans, and reports and documented in the requirements traceability documents. HSI requirements should also be entered and maintained in system requirements databases in the same manner as all other system requirements.
4. *Make HSI a Factor in Source Selection for Contracted Development Efforts* – HSI requirements must be explicit in source selection planning and implementation with adequate priority in the SOW.
    - HSI must be a distinctive major managerial and technical area requirement during source selection. Solicitations must require offerers to respond to all pertinent HSI considerations in the contract SOW.
    - HSI requirements should be included in contracting documents, such as the SOW, descriptions of deliveries and performance requirements, instructions to offerers, and evaluation factors for awarding the contract.
    - The source selection evaluation board should include an HSI representative(s) capable of adequately representing the HSI domains in source selection scoring.

5. *Execute Integrated Technical Processes* – HSI domain integration begins in early system concept development with front-end analyses and requirements definition and continues through development, operations, sustainment, modification, and all the way to eventual system disposal.
    - HSI activities and considerations must be included in each key project planning document (e.g., acquisition strategy, SEP, test plan, verification, etc.).
    - Systems engineers and HSI personnel must be prepared to present accurate, integrated cost data whenever possible to demonstrate reduced total ownership costs, thereby justifying trade-off decisions that may increase design and acquisition costs.
    - Program tools and processes should facilitate exchange of technical data between HSI activities, engineering and design disciplines, and the overall program.
    - HSI considerations should be included in the system architectural framework.
6. *Conduct Proactive Tradeoffs* – Often tradeoff decisions are made rapidly, so HSI representatives must be available and ready with information to influence HSI-related decisions.
    - The development of HSI objectives and thresholds in the requirements documents are critical to subsequent HSI tradeoff analyses.
    - In conducting tradeoff analyses both within HSI domains and for the system as a whole, the primary goal is to ensure human performance meets or exceeds the performance requirements for the total system without compromising survivability, environment, safety, occupational health, and habitability.
7. *Conduct* HSI Assessments – The purpose of the HSI assessment process is to evaluate the application of HSI principles throughout the system life cycle. The process enables cross-corporate and cross-discipline HSI collaboration in acquisition program evaluations and provides an integral avenue for HSI issue identification and resolution.
    - HSI assessment should be initiated early in the system development cycle and addressed throughout the system development process, particularly in SE technical reviews and

during working group meetings, design reviews, logistical assessments, verification, and validation

- HSI assessments should be based on sound data collection and analyses. Deficiencies should be documented and include a detailed description of the deficiency, its operational impact, recommended corrective action, and current status.

Some key web sites that will provide methods, tools, processes, referrals, and more detailed references to accomplish HSI in SE are as follows:

- <https://www.connect.incose.org/tb/specialty/hsi/default.aspx> – HSI Working Group site on INCOSE Connect
- <http://hsiport.resourceconsultants.com/register.aspx> – HSI Tools Port: Program Online Tools Support
- <http://www.dtic.mil/dticasd/> – DoD Defense Technical Information Center: Support Methods and Resources
- <http://www.paper-review.com/tools/tdb/home.php> – INCOSE Tools Database.

9.13 Value Engineering[21,22,23]

Value Engineering (VE), Value Management (VM), and Value Analysis (VA) are all terms that pertain to the application of the VE process. This chapter discusses how VE supports, compliments, and "adds value" to the SE discipline.

Value Engineering originated in studies of product changes resulting from material shortages during World War II. During that period, materials identified in designs were substituted without sacrificing quality and performance. Lawrence D. Miles, of General Electric (GE), noticed this success and developed a formal methodology to utilize teams to examine the function of products manufactured by GE.

Value Engineering uses a systematic process (i.e., formal job plan), VE-certified facilitators/team leads, and a multidisciplinary team approach to identify and evaluate solutions to complex problems in the life-cycle of a project, process, or system. The VE process utilizes several industry standard problem-solving/ decision-making techniques in an organized effort directed at independently analyzing the functions of programs, projects, organizations, processes, systems, equipment, facilities, services, and supplies. The objective is to achieve the essential functions at the lowest LCC consistent with required performance, reliability, availability, quality, and safety. VE is <u>not</u> a cost reduction activity, but a function-oriented method to improve the value of a product. There is no limit to the field in which VE may be applied.

The objective of performing VE is to improve the economical value of a project, product, or process by reviewing its elements to accomplish the following:

- Achieve the essential functions and requirements
- Lower total LCC (resources)
- Attain the required performance, safety, reliability, quality, etc.
- Meet schedule objectives.

Value is defined as a fair return or the equivalent in goods, services, or money for something exchanged. In other words, value is based on "what you get" relative to "what it cost." It is represented by the relationship:

$$\text{Value} = \frac{\text{Functions}}{\text{Resources}}$$

Performance is measured by the requirements from the customer. Cost is calculated in the materials, labor, price, time, etc., required to accomplish that function.

According to *The Value Standard*,[24] developed by the Society of American Value Engineers (SAVE), International, to qualify as a Value Study, the following conditions must be satisfied:

- The Value Study Team follows an organized, six-phase VE Job Plan (see below) and performs Function Analysis.
- The Value Study Team is a multidisciplinary group, chosen based on their expertise.
- The Value Team Leader (i.e., facilitator) is trained in the value methodology.

9.13.1 Systems Engineering Applicability

Value Engineering supports the various aspects and techniques of SE as well as being a comprehensive approach to SE project initiation. VE is implemented through a systematic, rational process consisting of a series of team-based techniques, including:

- Mission definition, strategic planning, or problem-solving techniques to determine current and future states. High-level requirements definition can be initiated in these processes.

- Function analysis to define what a system or system element does or its reason for existence. This technique serves as the basis or structure for technical, functional, and/or operational requirements.
- Innovative, creative, or speculative techniques to generate new alternatives. Trade studies are often defined in VE workshops from the more "viable" alternatives. Time can be given to conduct in-depth analyses, such as feasibility studies, cost estimates, etc. The VE workshop can then be reconvened (1-6 months later, for example) to determine the preferred alternative.
- Evaluation techniques to select preferred alternatives. These range in complexity from completely subjective to quantitative, depending on the level of rigor and justification required.

Value Engineering can be used to effectively and efficiently initiate an SE Project. The team approach brings the customer and stakeholders together along with engineering, planning, finance, marketing, etc. so that the strategic elements of the project can be discussed or developed. Customer needs can be identified and established as requirements. Functions are brainstormed and alternatives developed. In general, the boundaries of the scope of work are identified and clarified with all affected groups. The SE effort can then proceed with initial planning already established and customer input integrated. Some of the uses and benefits of VE include:

- Clarify objectives
- Solve problems (obtain "buy-in" to solution)
- Improve quality and performance
- Reduce costs and schedule
- Assure compliance
- Identify and evaluate additional concepts and options
- Streamline and validate processes and activities
- Strengthen teamwork
- Reduce risks
- Understand customer requirements.

9.13.2 VE Job Plan

The VE process uses a formal job plan (e.g., Systems Approach), which is a scientific method of problem solving that has been optimized over the last 50 years. The premise of VE is to use a beginning-to-end process that analyzes functions in such a way that creates change or value-added improvement in the end. A typical six-step VE Job Plan is as follows:

- *Phase 0: Preparation/Planning* – Plan the scope of the VE effort.
- *Phase 1: Information Gathering* – Clearly identify the problem(s) to be solved or the objective of the work scope. This includes gathering background information pertinent to the meeting objective and identifying and understanding customer/stakeholder requirements and needs.
- *Phase 2: Function Analysis* – Define the project functions using an active verb and measureable noun, then review and analyze the functions to determine which are critical, need improvements, or are unwanted. Several techniques can be used to enhance function analysis. Functions can be organized in a WBS; in a Function Analysis System Technique (FAST) Diagram (discussed below); or in a flow diagram.

  Functions can be assigned a cost. Depending on the objective of the VE Study, determining the cost/function relationships is one method to identify where unnecessary costs exist within the scope of the study. Other criteria to identify those areas needing improvement are personnel, environmental safety, quality, reliability, construction time, etc. Cost/Function relationships provide direction for the team related to areas that provide of greatest opportunity for cost improvement and the greatest benefits to the project. This can be documented in a Cost/Function Worksheet.
- *Phase 3: Creativity* – Brainstorm different ways to accomplish the function(s), especially those that are high-cost or low-value.
- *Phase 4: Evaluation* – Identify the most promising functions or concepts. The functions can also be used in this phase to generate full alternatives that accomplish the overall objective. Then the alternatives can be evaluated using the appropriate structured evaluation technique.
- *Phase 5: Development* – Develop the viable ideas into alternatives that are presented to decision makers to determine the path forward. Alternatively, the full alternatives that were evaluated in Phase 4 can be further developed into proposals, which may include rough-order-of-magnitude cost, estimated schedule, resources, etc.
- *Phase 6: Presentation / Implementation* – It may be desirable to present the alternatives/proposals to management or additional

customers for final direction. This is where the implementation plan concepts are identified and a point of contact assigned to manage the actions.

9.13.3 Function Analysis System Technique (FAST) Diagram

The FAST diagramming technique was developed in 1964 by Charles W. Bytheway to help identify the dependencies and relationships between functions. A FAST diagram (see Figure 9-6) is not time-oriented like a PERT chart or flow chart. It is a function-oriented model that applies intuitive logic to verify the functions that make up the critical path.

There are several different types of FAST diagrams as well as varying levels of complexity. The purpose is to develop a statement of the function (verb and noun) for each part or element of the item or process under study. Functions are classified as basic and secondary. The critical path is made up of the main functions within the scope of the activity or project. The Basic Function is typically the deliverable or end state of the effort. The Higher Order Function is the ultimate goal and is the answer to "why" the Basic Function is performed. Functions that "happen at the same time" or "are caused by" some function on the critical path are known as "When" functions and are placed below the critical path functions. Functions that happen "all the time," such as safety, aesthetic functions, etc., are placed above the critical path functions.

Unwanted Functions are noted by a double lined box. They are not essential to the performance of the scope, but are a consequence of the selected design solution. Limiting unwanted functions and minimizing the cost of basic/critical path functions results in an item of "best value" that is consistent with all performance, reliability, quality, maintainability, logistics support, and safety requirements.

There is no "correct" FAST model; team discussion and consensus on the final diagram is the goal. Using a team to develop the FAST diagram is beneficial for several reasons:

- Applies intuitive logic to verify functions
- Displays functions in a diagram or model
- Identifies dependence between functions
- Creates a common language for team
- Tests validity of functions.

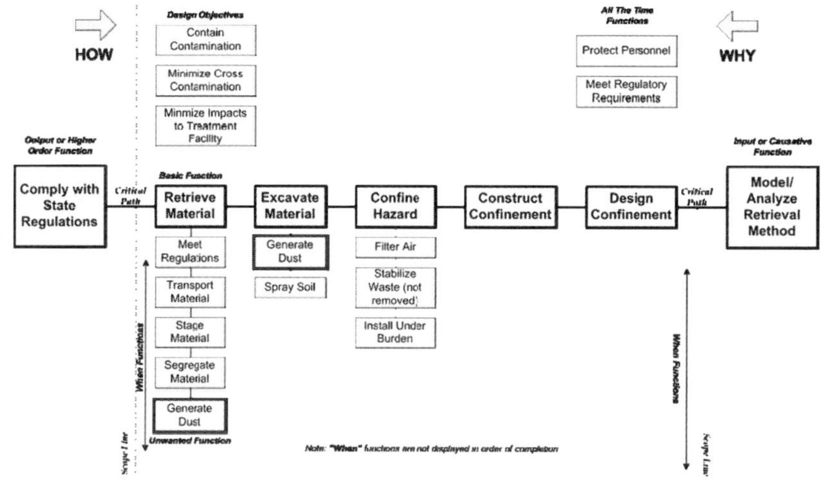

Figure 9-6 Sample Function Analysis System Technique (FAST) Diagram

9.13.4 Value Engineering Certification

A Certified Value Professional should be used to facilitate VE Workshops. They have the training and experience to manage a team, implement the methodology, and maximize the benefits to the customer.

The Certification Program is composed of two major elements: individual professional certification and educational program approval. The highest level is the *Certified Value Specialist* (CVS), which is recognition of the individual who has met all certification requirements, both technical and experience, and whose principal career is VE.

The *Associate Value Specialist* (AVS) program recognizes those individuals who have decided to become professional value engineers but who have not yet acquired all the experience or technical skills expected of a CVS. The *Value Methodology Practitioner* (VMP) program was established to recognize those individuals who acquired the basic skills of VE/VA but their principal career is not VE.

The CVS and VMP must recertify every four years. Although considered an entry level certification, the AVS may be maintained indefinitely as long as all certification maintenance fees are paid. Membership in SAVE is not a requirement for individual certification or for educational program approval.

9.13.5 Conclusion

Value Engineering is a best business practice. Projects that use VE in the early Development Stages have shown high rates of success. A VE study is more rigorous than the typical project review. Each VE Study brings together an impartial and independent team of technical experts with a common purpose of improving and optimizing the program or project's value. VE is seen as a systematic and creative approach for increasing the "return on investment" in systems, facilities, and other products. For more information on VE, consult the website of the Society of American Value Engineers, International.[23]

9.14 References

1. Blanchard, Ben, *Logistics Engineering and Management*, 5th Ed., Prentice Hall, 1998
2. Blanchard, Ben, and Wolter Fabrycky, *Systems Engineering and Analysis,* 3rd Ed., Prentice Hall, 1998
3. Ibid, Ch. 12-13 includes complete discussions of the metrics and calculations for Availability, Reliability and Maintainability
4. For discussion of Survivable Systems Analysis Methods see <http://www.cert.org/archive/html/analysis-method.htm>
5. Alexander, Ian, "Systems Engineering: -ilities for Victory," Downloaded from <http://easyweb.easynet.co.uk/iany/consultancy/systems_engineering/ilities_for_victory.htm>
6. see <http://www.washingtonwatchdog.org/documents/gao/05/GAO-05-966.html>
7. *Defense Acquisition Deskbook*, Version 3.1, 30 September 1999
8. The ISO 14000 family of International Standards, <http://www.iso.ch/iso/en/prods-services/otherpubs/iso14000/index.html>
9. Botkin, Daniel B., and Edward A. Keller, *Environmental Science: Earth as a Living Planet,* 2nd Ed., New York: John Wiley Sons, 1998
10. Edwards, Mary, an Assistant Professor at the University of Wisconsin, Madison, developed a guide that includes a chapter on environmental impact analysis. It can be found at <http://www.lic.wisc.edu/shapingdane/facilitation/all_resources/impacts/analysis_environmental.htm>
11. Fabrycky, W. J., and B. S. Blanchard, *Life-Cycle Cost and Economic Analysis*, Prentice Hall, 1991
12. "Life cycle cost analysis," (11 May 112006), In *Wikipedia, The Free Encyclopedia*. Retrieved May 22, 2006, from <http://en.wikipedia.org/w/index.php?title=Life_cycle_cost_analysis&oldid=52726603>
13. *Recommended Practices from the Society of Allied Weight Engineers*, <http://www.sawe.org>
14. U.S. Department of Labor, *Job Hazard Analysis*, Washington, DC: OSHA 3071, 2002 (available on-line at <http://www.osha.gov/Publications/osha3071.pdf>).
15. Bias, R. G., and D. J. Mayhew, *Cost Justifying Usability*, Boston, MA. Academic Press, 1994.
16. Blanchard and Fabrycky, *Systems Engineering and Analysis*, 4th Ed., N.J.: Pearson Prentice Hall, 2006
17. Booher, H. R. (Ed.), *Handbook of Human Systems Integration*, NY: J. Wiley & Sons, Inc., 2003
18. Chapanis, A., *Human Factors in Systems Engineering*, NY: J. Wiley & Sons, Inc., 1996
19. Rouse, W. B., *Design for Success: A Human-Centered Approach to Designing Successful Products and Systems*, New York: J. Wiley & Sons, Inc., 1991
20. U.S. Nuclear Regulatory Commission. *Fact Sheet on Three Mile Island Accident,* <http://www.nrc.gov/reading-rm/doc-collections/fact-sheets/3mile-isle.html>
21. Salvendy, Gavriel (Ed.), *Handbook of Industrial Engineering*, Purdue University, NY: J. Wiley & Sons, 1982
22. *Value Methodology Pocket Guide*, Data, GOAL/QPC
23. Society of American Value Engineers, International, August 2009, <http://www.value-eng.org/>
24. Society of American Value Engineers, International, *International Value Standard*, 2007 Ed.

Appendix A: System Life-Cycle Process N^2 Chart

Chart A-1 illustrates the input/output relationships between the various SE processes presented in the handbook. The individual processes are placed on the diagonal by abbreviation to the process names, as follows:

EXT – External inputs and outputs
SUP – Supply
ACQ - Acquisition
PP - Project Planning
PAC - Project Assessment and Control
DM - Decision Management
RM - Risk Management
CM - Configuration Management
INFOM - Information Management
MEAS - Measurement
SRD - Stakeholder Requirements Definition
RA - Requirements Analysis
AD - Architectural Design
IMPL - Implementation
INT - Integration
VER - Verification
TRAN - Transition
VAL - Validation
OPER - Operation
MAINT - Maintenance
DISP - Disposal
LCMM – Life Cycle Model Management
INFRAM - Infrastructure Management
PPM – Project Portfolio Management
HRM – Human Resource Management
QM - Quality Management
TLR – Tailoring.

The rest of the squares represent the inputs/outputs interface shared by the processes that intersect at a given square. Outputs flow horizontally; input flow vertically.

Chart A-1. System Life-Cycle Process N^2 chart per the INCOSE SE Handbook

EXT	x	x	x							x											x		x		x	x
x	SUP		x	x	x	x	x	x															x			x
x		ACQ	x	x	x	x	x	x						x												x
	x	x	PP	x	x	x	x	x	x	x	x	x	x	x	x	x	x	x	x	x		x		x	x	x
	x	x	x	PAC	x	x	x	x	x	x	x	x	x	x	x	x	x	x	x	x			x	x	x	x
			x	x	DM	x	x	x																		
			x	x	x	RM	x	x																		
			x	x	x	x	CM	x																		
x			x	x	x	x	x	INFOM																		
x			x	x	x	x	x	x	MEAS												x					
					x	x	x	x	x	SRD	x	x	x	x	x	x	x	x	x	x						
					x	x	x	x	x		RA	x	x		x											
					x	x	x	x	x			AD	x	x	x											
		x	x		x	x	x	x				x	IMPL	x		x		x	x							
		x	x	x	x	x	x	x				x		INT	x	x										
		x	x	x	x	x	x	x				x			VER	x	x									
		x	x	x	x	x	x	x				x				TRAN	x									
		x	x	x	x	x	x	x				x					VAL	x	x	x						
		x	x	x	x	x	x	x				x						OPER	x							
		x	x	x	x	x	x	x				x							MAINT	x						
x		x	x	x	x	x	x	x				x								DISP						
			x		x	x	x	x	x												LCMM				x	
				x	x	x	x	x													x	INFRAM				
	x		x		x	x	x	x													x	x	PPM	x	x	x
x			x		x	x	x	x																HRM		
x			x		x	x	x	x													x				QM	
					x	x	x	x													x					TLR

Note: Absence of an x in an intersection does not preclude tailoring to create a relationship between any two processes.

Appendix B: System Life-Cycle Process Mappings

Appendix B provides top-level mappings between some of the key, industry-accepted SE standards (or de facto standards) and the various SE processes presented in this handbook. These mappings are provided as a resource and should not be interpreted as a substitute for obtaining and reviewing the actual standards.

B.1 ISO-15288:2002 Mapping

ISO/IEC 15288:2002 System Life Cycle Process	INCOSE SE Handbook v3.2.1 Section	Notes/Comments
6.1 Agreement Processes		
6.1.1 Acquisition Process	6.1 Acquisition Process	
6.1.2 Supply Process	6.2 Supply Process	
6.2 Enterprise Processes		
6.2.1 Enterprise Environment Management Process	7.1 Life Cycle Model Management Process	Policies & procedures
6.2.2 Investment Management Process	7.3 Project Portfolio Management Process	
6.2.3 System Life Cycle Process Mgmt Process	7.1 Life Cycle Model Management Process 7.2 Infrastructure Management Process 5.7 Measurement Process	Methods & tools Enterprise measures
6.2.4 Resource Management Process	7.2 Infrastructure Management Process 7.4 Human Resource Management Process	Includes all types of resources
6.2.5 Quality Management Process	7.5 Quality Management Process	
6.3 Project Processes		
6.3.1 Project Planning Process	5.1 Project Planning Process	
6.3.2 Project Assessment Process	5.2 Project Assessment and Control Process 5.7 Measurement Process	Project status Project measures
6.3.3 Project Control Process	5.2 Project Assessment and Control Process	Direction & corrective actions
6.3.4 Decision-Making Process	5.3 Decision Management Process	
6.3.5 Risk Management Process	5.4 Risk Management Process	
6.3.6 Configuration Management Process	5.5 Configuration Management Process	
6.3.7 Information Management Process	5.6 Information Management Process	
6.4 Technical Processes		
6.4.1 Stakeholder Requirements Definition Process	4.1 Stakeholder Requirements Definition Process	

ISO/IEC 15288:2002 System Life Cycle Process	INCOSE SE Handbook v3.2.1 Section	Notes/Comments
6.4.2 Requirements Analysis Process	4.2 Requirements Analysis Process	
6.4.3 Architectural Design Process	4.3 Architectural Design Process	
6.4.4 Implementation Process	4.4 Implementation Process	
6.4.5 Integration Process	4.5 Integration Process	
6.4.6 Verification Process	4.6 Verification Process	
6.4.7 Transition Process	4.7 Transition Process	
6.4.8 Validation Process	4.8 Validation Process	
6.4.9 Operation Process	4.9 Operation Process	
6.4.10 Maintenance Process	4.10 Maintenance Process	
6.4.11 Disposal Process	4.11 Disposal Process	
A Tailoring Process		
A.2 Tailoring Process	8.1 Tailoring Process	

B.2 ISO-15288:2008 Mapping

ISO/IEC 15288:2008 System Life Cycle Process	INCOSE SE Handbook v3.2.1 Section	Notes/Comments
6.1 Agreement Processes		
6.1.1 Acquisition Process	6.1 Acquisition Process	
6.1.2 Supply Process	6.2 Supply Process	
6.2 Organizational Project-Enabling Processes		
6.2.1 Life Cycle Model Management Process	7.1 Life Cycle Model Management Process	
6.2.2 Infrastructure Management Process	7.2 Infrastructure Management Process	
6.2.3 Project Portfolio Management Process	7.3 Project Portfolio Management Process	
6.2.4 Human Resource Management Process	7.4 Human Resource Management Process	
6.2.5 Quality Management Process	7.5 Quality Management Process	
6.3 Project Processes		
6.3.1 Project Planning Process	5.1 Project Planning Process	
6.3.2 Project Assessment and Control Process	5.2 Project Assessment and Control Process	
6.3.3 Decision Management Process	5.3 Decision Management Process	
6.3.4 Risk Management Process	5.4 Risk Management Process	
6.3.5 Configuration Management Process	5.5 Configuration Management Process	
6.3.6 Information Management Process	5.6 Information Management Process	
6.3.7 Measurement Process	5.7 Measurement Process	

ISO/IEC 15288:2008 System Life Cycle Process	INCOSE SE Handbook v3.2.1 Section	Notes/Comments
6.4 Technical Processes		
6.4.1 Stakeholder Requirements Definition Process	4.1 Stakeholder Requirements Definition Process	
6.4.2 Requirements Analysis Process	4.2 Requirements Analysis Process	
6.4.3 Architectural Design Process	4.3 Architectural Design Process	
6.4.4 Implementation Process	4.4 Implementation Process	
6.4.5 Integration Process	4.5 Integration Process	
6.4.6 Verification Process	4.6 Verification Process	
6.4.7 Transition Process	4.7 Transition Process	
6.4.8 Validation Process	4.8 Validation Process	
6.4.9 Operation Process	4.9 Operation Process	
6.4.10 Maintenance Process	4.10 Maintenance Process	
6.4.11 Disposal Process	4.11 Disposal Process	
A Tailoring Process		
A.2 Tailoring Process	8.1 Tailoring Process	

B.3 ISO/IEC 26702:2007/IEEE 1220 ™-2005 Mapping

ISO/IEC 26702:2007/IEEE 1220 ™-2005 Process	INCOSE SE Handbook v3.2.1 Section	Notes/Comments
4. General requirements		
4.1 Systems engineering process	3 Generic Life-Cycle Stages 7.1 Life Cycle Model Management Process	
4.2 Policies and procedures for systems engineering	7.1 Life Cycle Model Management Process	
4.3 Planning the technical effort	5.1 Project Planning Process	
4.4 Development strategies	3 Generic Life-Cycle Stages 7.1 Life Cycle Model Management Process 5.1 Project Planning Process	
4.5 Modeling and prototyping	4.12.1 Modeling, Simulation, and Prototyping 4.2 Requirements Analysis Process 4.3 Architectural Design Process 4.6 Verification Process 4.8 Validation Process	
4.6 Integrated repository	5.6 Information Management Process	

ISO/IEC 26702:2007/IEEE 1220 ™-2005 Process	INCOSE SE Handbook v3.2.1 Section	Notes/Comments
4.7 Integrated data package	4.3 Architectural Design Process 4.4 Implementation Process 5.6 Information Management Process	
4.8 Specification tree	4.2 Requirements Analysis Process 4.3 Architectural Design Process	
4.9 Drawing tree	4.2 Requirements Analysis Process 4.3 Architectural Design Process	
4.10 System breakdown structure	4.3 Architectural Design Process	
4.11 Integration of the systems engineering effort	5.1 Project Planning Process 5.2 Project Assessment and Control Process 7.4 Human Resource Management Process	
4.12 Technical reviews	5.2 Project Assessment and Control Process	
4.13 Quality management	7.5 Quality Management Process	
4.14 Product and process improvement	7.1 Life Cycle Model Management Process 7.5 Quality Management Process	
5. Application of systems engineering throughout the system life cycle		
5.1 System definition stage	3 Generic Life-Cycle Stages	Different stages
5.2 Preliminary design stage	3 Generic Life-Cycle Stages	Different stages
5.3 Detailed design stage	3 Generic Life-Cycle Stages	Different stages
5.4 Fabrication, assembly, integration, and test stage	3 Generic Life-Cycle Stages	Different stages
5.5 Production and Support Stages	3 Generic Life-Cycle Stages	Different stages
5.6 Simultaneous engineering of life cycle processes	3 Generic Life-Cycle Stages	
6. The systems engineering process		
6.1 Requirements analysis	4.1 Stakeholder Requirements Definition Process 4.2 Requirements Analysis Process	
6.2 Requirements validation	4.1 Stakeholder Requirements Definition Process 4.2 Requirements Analysis Process 4.8 Validation Process	Special emphasis on early validation

ISO/IEC 26702:2007/IEEE 1220 ™-2005 Process	INCOSE SE Handbook v3.2.1 Section	Notes/Comments
6.3 Functional analysis	4.2 Requirements Analysis Process 4.8 Validation Process	
6.4 Functional verification	4.2 Requirements Analysis Process 4.6 Verification Process	Special emphasis on early verification
6.5 Synthesis	4.3 Architectural Design Process	
6.6 Design verification	4.3 Architectural Design Process 4.6 Verification Process	Special emphasis on early verification
6.7 Systems analysis	5.3 Decision Management Process 5.4 Risk Management Process 4.2 Requirements Analysis Process 4.3 Architectural Design Process	
6.8 Control	5.2 Project Assessment and Control Process 5.4 Risk Management Process 5.5 Configuration Management Process 5.6 Information Management Process 4.3 Architectural Design Process	

B.4 ANSI EIA-632 (1999) Mapping

ANSI-EIA-632-1999 Requirements	INCOSE SE Handbook v3.2.1 Section	Notes/Comments
Supply Process		
1 - Product Supply	6.2 Supply Process	
Acquisition Process		
2 - Product Acquisition	6.1 Acquisition Process	
3 - Supplier Performance	6.1 Acquisition Process	
Planning Process		
4 - Process Implementation Strategy	7.1 Life Cycle Model Management Process	
5 - Technical Effort Definition	5.1 Project Planning Process	
6 - Schedule and Organization	5.1 Project Planning Process	
7 - Technical Plans	5.1 Project Planning Process	
8 - Work Directives	5.1 Project Planning Process	
Assessment Process		
9 - Progress Against Plans and Schedules	5.2 Project Assessment and Control Process	
10 - Progress Against Requirements	5.2 Project Assessment and Control Process	

ANSI-EIA-632-1999 Requirements	INCOSE SE Handbook v3.2.1 Section	Notes/Comments
11 - Technical Reviews	5.2 Project Assessment and Control Process	
Control Process		
12 - Outcomes Management	5.2 Project Assessment and Control Process 5.5 Configuration Management Process 5.4 Risk Management Process 5.6 Information Management Process	
13 - Information Dissemination	5.6 Information Management Process	
Requirements Definition		
14 - Acquirer Requirements	4.1 Stakeholder Requirements Definition Process	
15 - Other Stakeholder Requirements	4.1 Stakeholder Requirements Definition Process	
16 - System Technical Requirements	4.2 Requirements Analysis Process	
Solution Definition		
17 - Logical Solution Representations	4.3 Architectural Design Process	
18 - Physical Solution Representations	4.3 Architectural Design Process	
19 - Specified Requirements	4.3 Architectural Design Process	
Implementation Process		
20 - Implementation	4.4 Implementation Process 4.5 Integration Process	
Transition to Use Process		
21 - Transition to Use	4.7 Transition Process	
System Analysis		
22 - Effectiveness Analysis	9 Specialty Engineering Activities	
23 - Tradeoff Analysis	5.3 Decision Management Process	
24 - Risk Analysis	5.4 Risk Management Process	
Requirement Validation		
25 - Requirements Statements Validation	4.2 Requirements Analysis Process 4.8 Validation Process	Special emphasis on early validation
26 - Acquirer Requirements Validation	4.1 Stakeholder Requirements Definition Process 4.8 Validation Process	Special emphasis on early validation
27 - Other Stakeholder Requirements Validation	4.1 Stakeholder Requirements Definition Process 4.8 Validation Process	Special emphasis on early validation
28 - System Technical Requirements Validation	4.3 Architectural Design Process 4.8 Validation Process	Special emphasis on early validation

ANSI-EIA-632-1999 Requirements	INCOSE SE Handbook v3.2.1 Section	Notes/Comments
29 - Logical Solution Representations Validation	4.3 Architectural Design Process 4.8 Validation Process	Special emphasis on early validation
System Verification Process		
30- Design Solution Verification	4.3 Architectural Design Process 4.6 Verification Process	Special emphasis on early verification
31- End Product Verification	4.6 Verification Process	
32 - Enabling Product Readiness	4.6 Verification Process	
End Products Validation		
33 - End Products Validation	4.8 Validation Process	

B.5 CMMI®-for Development v1.2 Mapping

CMMI® for Development V1.2 Process Areas	INCOSE SE Handbook v3.2.1 Section	Notes/Comments
Process Management		
Organizational Process Focus (OPF)	7.1 Life Cycle Model Management Process	
Organizational Process Definition+IPPD (OPD+IPPD)	7.1 Life Cycle Model Management Process 7.2 Infrastructure Management Process 8.1 Tailoring Process	
Organizational Training (OT)	7.4 Human Resource Management Process	
Organizational Process Performance (OPP)	7.1 Life Cycle Model Management Process 5.7 Measurement Process	Limited coverage in the SE Handbook
Organizational Innovation and Deployment (OID)	7.1 Life Cycle Model Management Process	
Project Management		
Project Planning (PP)	5.1 Project Planning Process	
Project Monitoring and Control (PMC)	5.2 Project Assessment and Control Process	
Supplier Agreement Management (SAM)	6.1 Acquisition Process	
Integrated Project Management+IPPD (IPM+IPPD)	5.1 Project Planning Process 5.2 Project Assessment and Control Process 8.1 Tailoring Process 4.1 Stakeholder Requirements Definition Process	
Risk Management (RSKM)	5.4 Risk Management Process	
Quantitative Project Management (QPM)	5.1 Project Planning Process 5.2 Project Assessment and Control Process 5.7 Measurement Process	Limited coverage in the SE Handbook

CMMI® for Development V1.2 Process Areas	INCOSE SE Handbook v3.2.1 Section	Notes/Comments
Engineering		
Requirements Development (RD)	4.1 Stakeholder Requirements Definition Process 4.2 Requirements Analysis Process	Deals with customer, product, and product component requirements
Requirements Management (REQM)	4.1 Stakeholder Requirements Definition Process 4.2 Requirements Analysis Process 5.2 Project Assessment and Control Process	
Technical Solution (TS)	4.3 Architectural Design Process 4.4 Implementation Process	
Product Integration (PI)	4.5 Integration Process 4.7 Transition Process	
Verification (VER)	4.6 Verification Process	Includes peer reviews
Validation (VAL)	4.8 Validation Process	
Support		
Configuration Management (CM)	5.5 Configuration Management Process 5.6 Information Management Process	
Process and Product Quality Assurance (PPQA)	7.5 Quality Management Process	
Measurement and Analysis (MA)	5.7 Measurement Process	
Decision Analysis and Resolution (DAR)	5.3 Decision Management Process	
Causal Analysis and Resolution (CAR)	5.2 Project Assessment and Control Process 5.7 Measurement Process	Limited coverage in the SE Handbook

Appendix C: Acronym List

AECL	Atomic Energy Commission Limited [Canada]
AHP	Analytic Hierarchy Process
ANSI	American National Standards Institute
AVS	Associate Value Specialist
CCB	Configuration Control Board
CDRL	Contract Data/Document Requirements List
CER	Cost Estimating Relationship
Cf	consequence of failure
CI	configuration item
CMMI®	Capability Maturity Model® Integration
CMP	Configuration Management Plan
ConOps	Concept of Operations
COTS	Commercial Off-the-Shelf
CSCI	Computer Software Configuration Item
CSEP	Certified Systems Engineering Professional
CVS	Certified Value Specialist
DAU	Defense Acquisition University
DOD	Department of Defense [USA]
DODAF	Department of Defense Architecture Framework [USA]
ECP	Engineering Change Proposal
ECR	Engineering Change Request
EIA	Electronic Industries Alliance
EMC	ElectroMagnetic Compatibility
EMI	ElectroMagnetic Interference
ESEP	Expert Systems Engineering Professional
FAST	Function Analysis System Technique
FFBD	Functional Flow Block Diagram
FMECA	Failure Modes, Effects, and Criticality Analysis
G&A	General and Administrative
GAO	Government Accountability Office [USA]
GE	General Electric
GNP	Gross National Product
HFE	Human Factors Engineering
HSI	Human Systems Integration
HWCI	Hardware Configuration Item

ICD	Interface Control Document
IDEF	Integrated Definition for Functional Modeling
IEC	International Electrotechnical Commission
IEEE	Institute of Electrical and Electronics Engineers
IFWG	Interface Working Group
IID	Incremental and Iterative Development
ILS	Integrated Logistics Support
INCOSE	International Council on Systems Engineering
IPD	Integrated Product Development
IPDT	Integrated Product Development Team
IPPD	Integrated Product and Process Development
IPT	Integrated Product Team
ISO	International Organization for Standardization
IV&V	integration, verification, and validation
KTA	Kepner-Tregoe Analysis
LAI	Lean Advancement Initiative
LCC	life-cycle cost
LEfSE	Lean Enablers for Systems Engineering
LINAC	linear accelerator
LSA	Logistic Support Analysis
MAUA	Multi-Attribute Utility Analysis
MBSE	Model-Based Systems Engineering
MEU	Maximum Expected Utility
MIT	Massachusetts Institute of Technology
MODAF	Ministry of Defense Architecture Framework [UK]
MOE	Measure of Effectiveness
MOP	Measure of Performance
MPE	Mass Properties Engineering
NASA	National Aeronautics and Space Administration [USA]
NCOSE	National Council on Systems Engineering (pre-1995)
NDI	Non-Developmental Item
O&S	Operations and Support
OCD	Operational Concept Document, see *ConOps*
OMG	Object Management Group
OOSEM	Object-Oriented Systems Engineering Method
OpEMCSS	Operational Evaluation Modeling for Context-Sensitive Systems
OSI	Open System Interconnect (for communication protocols)

PBS	Product Breakdown Structure
PCR	Process Compliance Review
PDT	Product Development Team
PERT	Program Evaluation Review Technique
Pf	probability of failure
PHS&T	Packaging, Handling, Storage, and Transportation
PIT	Product Integration Team
PLCS	Product Life-cycle Support
PSM	Practical Software and Systems Measurement
PUID	Project Unique Identifier
QA	Quality Assurance
QFD	Quality Function Deployment
QMP	Quality Management Plan
R&D	research and development
RCM	Reliability Centered Maintenance
RFP	Request for Proposal
RMP	Risk Management Plan
RTM	Requirements Traceability Matrix
RVTM	Requirements Verification and Traceability Matrix
SAVE	Society of American Value Engineers
SCN	Specification Change Notice
SE	Systems Engineering
SEARI	Systems Engineering Advancement Research Institute
SEH	Systems Engineering Handbook
SEIT	Systems Engineering and Integration Team
SEMP	Systems Engineering Management Plan
SEMS	Systems Engineering Master Schedule
SEP	Systems Engineering Plan
SoS	system-of-systems
SOW	Statement of Work
SRD	System Requirements Document
SRR	System Requirements Review
STEP	Standard for the Exchange of Product Model Data
SWOT	Strength-Weakness-Opportunity-Threat
SysML™	Systems Modeling Language
SYSPG	Systems Engineering Process Group
TBD	to be determined

TBR	to be reviewed
TBS	to be supplied
TPM	Technical Performance Measure
TQM	Total Quality Management
TR	Technical Report
U.S.	United States
UK	United Kingdom
UL	Underwriters Laboratory [USA and Canada]
UML™	Unified Modeling Language™
USA	United States of America
USD	U.S. dollars
V&V	verification and validation
VA	Value Analysis
VE	Value Engineering
VM	Value Management
VMP	Value Methodology Practitioner
WBS	Work Breakdown Structure
WG	Working Group

Appendix D: Terms and definitions

The term and definitions *in italic font style* are from ISO/IEC 15288:2008 – *Systems engineering – System life-cycle processes.*

Words not included in this glossary carry meanings consistent with dictionary definitions.

acquirer	*the stakeholder that acquires or procures a product or service from a supplier*
Acquisition logistics	Technical and management activities conducted to ensure supportability implications are considered early and throughout the Acquisition Process to minimize support costs and to provide the user with the resources to sustain the system in the field
Activity	*a set of cohesive tasks of a process*
Agile	Project execution methods can be described on a continuum from "adaptive" to "predictive." Agile methods exist on the "adaptive" side of this continuum, which is not the same as saying that agile methods are "unplanned" or "undisciplined."
Agreement	*the mutual acknowledgement of terms and conditions under which a working relationship is conducted*
Assembly	A system element comprising an integrated set of components and/or subassemblies that form a defined portion of a subsystem
Baseline	The gate-controlled step-by-step elaboration of business, budget, functional, performance, and physical characteristics, mutually agreed to by buyer and seller, and under formal change control. Baselines can be modified between formal decision gates by mutual consent through the change control process.
Capability	An expression of a system, product, function or process' ability to achieve a specific objective under stated conditions.
Commercial off-the-shelf (cots)	Commercial items that require no unique acquirer modifications or maintenance over the life cycle of the product to meet the needs of the procuring agency
Component	A system element comprised of multiple parts; a cleanly identified item.
Configuration	A characteristic of a system element, or project artifact, describing their maturity or performance.

Configuration item (CI)	A hardware, software, or composite item at any level in the system hierarchy designated for configuration management. (The system and each of its elements are individual CIs.) CIs have four common characteristics: 1. Defined functionality, 2. Replaceable as an entity, 3. Unique specification, 4. Formal control of form, fit, and function
Context diagram	This version of the handbook provides a high level view of the process-of-interest. The diagram summarizes the process activities and their inputs and outputs from/to external actors; some inputs are categorized as controls and enablers. A control governs the accomplishments of the process; an enabler is the means by which the process is performed.
Decision gate	A decision gate is an approval event (often associated with a review meeting). Entry and exit criteria are established for each decision gate; continuation beyond the decision gate is contingent on the agreement of decision-makers.
Derived requirements	Detailed characteristics of the system-of interest that typically are identified during elicitation of stakeholder requirements, requirements analysis, trade studies or validation
Design constraints	The boundary conditions, externally or internally imposed, for the system-of-interest within which the organization must remain when executing the processes during the concept and Development Stage
Element	*See System element*
Enabling system	*a system that supports a system-of-interest during its life-cycle stages but does not necessarily contribute directly to its function during operation*
Enterprise	*See Organization*
Environment	The surroundings (natural or man-made) in which the system-of-interest is utilized and supported; or in which the system is being developed, produced and retired.
Facility	*the physical means or equipment for facilitating the performance of an action, e.g. buildings, instruments, tools*
Failure	The event in which any part of an item does not perform as required by its specification. The failure may occur at a value in excess of the minimum required in the specification, i.e., past design limits or beyond the margin of safety.

Human factors	The systematic application of relevant information about human abilities, characteristics, behavior, motivation, and performance. It includes principles and applications in the areas of human related engineering, anthropometrics, ergonomics, job performance skills and aids, and human performance evaluation.
Human systems integration	The interdisciplinary technical and management processes for integrating human considerations within and across all system elements; an essential enabler to SE practice.
"-ilities"	The developmental, operational, and support requirements a program must address (e.g., availability, maintainability, vulnerability, reliability, supportability, etc.).
Integration definition for Functional Modeling (IDEF)	A family of modeling languages in the fields of systems and software engineering that provide a multiple page (view) model of a system that depicts functions and information or product flow. Boxes illustrate functions and arrows illustrate information and product flow. Alphanumeric coding is used to denote the view. IDEF0 - functional modeling method IDEF1 - information modeling method IDEF1X - data modeling method IDEF3 - process description capture method IDEF4 - object oriented design method IDEF5 - ontology description capture method
Life-cycle cost (LCC)	The total cost to the organization of acquisition and ownership of a system over its entire life. It includes all costs associated with the system and its use in the concept, development, production, utilization, support and Retirement Stages.
Life-Cycle model	*a framework of processes and activities concerned with the life cycle, which also acts as a common reference for communication and understanding*
Measure of Effectiveness	A measure used to quantify the performance of a system, product or process in terms that describe a measure to what degree the real objective is achieved.
N^2 diagrams	This graphical representation can be used to define the internal operational relationships or external interfaces of the system-of-interest.
Operator	*an individual who, or an organization that, contributes to the functionality of a system and draws on knowledge, skills and procedures to contribute the function*
Organization	*person or a group of people and facilities with an arrangement of responsibilities, authorities and relationships [adapted from ISO 9000:2005]*
Performance	A quantitative measure characterizing a physical or functional attribute relating to the execution of a process, function, activity or task; Performance attributes include quantity (how many or how much), quality (how well), timeliness (how responsive, how frequent), and readiness (when, under which circumstances).

Process	*set of interrelated or interacting activities which transforms inputs into outputs [ISO 9000:2000]*
Project	*an endeavor with defined start and finish criteria undertaken to create a product or service in accordance with specified resources and requirements*
Proof-of-concept	A naïve realization of an idea or technology to demonstrate its feasibility
Requirement	A statement that identifies a system, product or process' characteristic or constraint, which is unambiguous, clear, unique, consistent, stand-alone (not grouped), and verifiable, and is deemed necessary for stakeholder acceptability.
Resource	*an asset that is utilized or consumed during the execution of a process*
Specialty engineering	Analysis of specific features of a system that requires special skills to identify requirements and assess their impact on the system life cycle
Stage	*a period within the life cycle of an entity that relates to the state of its description or realization*
Stakeholder	*a party having a right, share or claim in a system or in its possession of characteristics that meet that party's needs and expectations*
Subassembly	A system element comprising an integrated set of components and/or parts that comprise a well-defined portion of an assembly, e.g. a video display with its related integrated circuitry or a pilot's radio headset.
Subsystem	A system element comprising an integrated set of assemblies, which performs a cleanly and clearly separated function, involving similar technical skills, or a separate supplier. Examples are an aircraft on-board communications subsystem or an airport control tower as a subsystem of the air transportation system.
Supplier	*an organization or an individual that enters into an agreement with the acquirer for the supply of a product or service*
System	*a combination of interacting elements organized to achieve one or more stated purposes* an integrated set of elements, subsystems, or assemblies that accomplish a defined objective. These elements include products (hardware, software, firmware), processes, people, information, techniques, facilities, services, and other support elements. (INCOSE) An example would be an air transportation system.
System element	*member of a set of elements that constitutes a system* a major product, service, or facility of the system, e.g. the aircraft element of an air transportation system (the term subsystem is sometimes used instead of element)

System life cycle	*the evolution with time of a system-of-interest from conception through to retirement*
System-of-interest	*the system whose life cycle is under consideration*
System-of-systems	System-of-systems applies to a system-of-interest whose system elements are themselves systems; typically these entail large scale inter-disciplinary problems with multiple, heterogeneous, distributed systems.
Systems engineering	Systems Engineering (SE) is an interdisciplinary approach and means to enable the realization of successful systems. It focuses on defining customer needs and required functionality early in the development cycle, documenting requirements, and then proceeding with design synthesis and system validation while considering the complete problem: operations, cost and schedule, performance, training and support, test, manufacturing, and disposal. SE considers both the business and the technical needs of all customers with the goal of providing a quality product that meets the user needs. [INCOSE]
Systems engineering effort	Systems engineering effort integrates multiple disciplines and specialty groups into a set of activities that proceed from concept to production to operation. SE considers both the business and the technical needs of all stakeholders with the goal of providing a quality system that meets their needs.
Systems engineering plan	Structured information describing how the system engineering effort, in form of tailored processes and activities, for one or more life-cycle stages, will be managed and conducted in the organization for the actual project.
Tailoring	The manner in which any selected issue is addressed in a particular project. The organization may seek to minimize the time and efforts it takes to satisfy an identified need consistent with common sense, sound business management practice, applicable laws and regulations, and the time sensitive nature of the requirement itself. Tailoring may be applied to various aspects of the project, including project documentation, processes and activities performed in each life-cycle stage, the time and scope of reviews, analysis, and decision-making consistent with all applicable statutory requirements.
Trade-off	*decision-making actions that select from various requirements and alternative solutions on the basis of net benefit to the stakeholders*
User	*individual who or group that benefits from a system during its utilization*
Validation	*confirmation, through the provision of objective evidence, that the requirements for a specific intended use or application have been fulfilled [ISO 9000: 2000]*
Verification	*confirmation, through the provision of objective evidence, that specified requirements have been fulfilled [ISO 9000: 2000]*

This page intentionally left blank.

Appendix E: Acknowledgements

SEHv3.2.1 Contributions

The primary purpose of this version 3.2.1 update was to incorporate comments received from ISO/IEC JTC1/SC7 member countries as a result of their consideration of the INCOSE Systems Engineering Handbook as an ISO/IEC Technical Report. Approximately 50 comments were received from the national bodies under the auspices of ISO/IEC JTC1/SC7. The editors would specifically like to thank the JTC1/SC7 member countries for their diligent review of the handbook; namely Australia, Canada (SCC), Great Britain, Japan, South Africa, and the United States (ANSI). Those comments significantly improved the quality and alignment of this handbook with ISO/IEC 15288:2008. In addition, members of ISO/IEC JTC1/SC7/WG20, under the leadership of Juan Garbosa; and the co-editors, Hironori Washizaki and Kevin Forsberg (ESEP), have been instrumental in the creation and evolution of the Technical Report of this handbook.

This update also corrected several minor issues identified by the INCOSE community. We would also like to thank the specific and anonymous reviewers who provided comments on v3.2.

INCOSE Technical Operations review and approval of version 3.2.1 was received on [TBD] with formal acceptance of Handbook v3.2.1 as an ISO/IEC Technical Report on [TBD]. [TBD] led this INCOSE review, and their efforts are much appreciated.

Gratefully,

Mike Krueger, CSEP

Dave Walden, CSEP

Douglas Hamelin

SEHv3.2 Contributions

The editorial team owes a debt of gratitude to those individuals and groups who contributed to previous versions of the *INCOSE Systems Engineering Handbook*. While some new material has been added, the majority of the effort to create this version was the rearrangement of existing material.

The primary purpose of this version 3.2 update was to: bring the text into alignment with ISO/IEC 15288:2008(E); resolve inconsistencies in version 3.1; consolidate related process information throughout the text to remove the multiple treatment of topics; and minimize impact to the INCOSE Certified Systems Engineering Professional (CSEP) examination. The editors wish to thank Vitech Corporation for the use of their CORE tool to help ensure consistency in the handbook context diagrams.

Several topics were added or updated from version 3.1. The topics, in alphabetical order, and the primary authors and Working Groups (WGs) are as follows: *Agile Development*, Kevin Forsberg (ESEP) and Hal Mooz (CSEP); *Lean Development*, Deb Secor and the INCOSE Lean Systems Engineering WG; *Measurement*, Paul Frenz (CSEP) and the INCOSE Measurement WG; *Modeling*, Jack Ring; *Object-Oriented Systems Engineering Method*, Jeff Estephan and the INCOSE Model-Based Systems Engineering (MBSE) WG; *Systems Modeling Language*, Sandy Friedenthal and the MBSE WG; *Usability/Human-System Integration*, Mike Mueller and the INCOSE HSI WG; *Value Engineering*, Lori Braase and the Idaho National Laboratory Value Engineering Group.

The review team was led by Yoshi Ohkami (ESEP), Dick Wray and G. Richard Freeman (CSEP) and included, in alphabetical order: Stu Allison, Samantha Brown, James Cademartori (CSEP), John Clark (CSEP), Kevin Forsberg (ESEP), , Cheryl Jones, Troy Petersen (CSEP), Karen Richter, Garry Roedler (ESEP), Seiko Shirasaka, Pete Suthon (CSEP), L. Mark Walker (CSEP), Hironori Washizaki, and Mike Zabat (CSEP). This team provided over 300 comments that significantly improved the quality of this handbook.

The editors also wish to acknowledge the Idaho National Laboratory Systems Engineering Department for their significant support of this update and thank them for their contribution of technical editor Douglas Hamelin.

Gratefully,

Mike Krueger, CSEP

Dave Walden, CSEP

SEHv3.1 Contributions

The development team for the *INCOSE Systems Engineering Handbook* version 3.1 owes a debt of gratitude to the people who contributed to versions 2a and 3.0. This update merges key elements of both of those activities. Everything in this update is based on those two solid platforms.

One of the constraints imposed on the development of Systems Engineering Handbook version 3.0 (SEHv3) was a page count limit of 150 pages. There were detailed sections in SEHv2a that we considered relevant, but which could not be included because of page count limitations. The intent in 2006 was that these sections would be extracted and put into an electronic file as a supplement to SEHv3. This revision is the initial step toward building a set of "living" appendices providing elaboration of the SE practices. We expect these appendices to evolve over time.

An important function of this handbook is to provide a body of knowledge for the INCOSE certification examination for Certified Systems Engineering Professionals (CSEP). Certification exams based on SEHv3.1 will include questions based on the information found in the appendices. As the body of knowledge expands with additional appendices and deletions of old material, exam questions will be deleted, modified, and/or added. Ample notification of such changes will be provided.

Several INCOSE working groups provided valuable contributions to this document, and we would like to specifically thank the Human Systems Integration WG, the Requirements WG, and the System Safety WG for their interest and support.

A review team, under the direction of Erik Aslaksen (CSEP), provided an invaluable assessment, resulting in 249 suggestions or comments, all of which were addressed. We would like to thank the entire team for their excellent review efforts: Dick Allen-Shalless, Eileen Arnold (CSEP), Gary Bakken, Carlos Caldeira, Mark Halverson, Jerry Huller (CSEP), Ken Kepchar (CSEP), Leonid Lev, Bernard Morais, John Muehlbauer (CSEP), Ramakrishnan Raman (CSEP), Alex Schmarr, and Mark Walker (CSEP). In addition, we would like to give especial thanks to Jerry Huller (CSEP) for his added effort in providing insightful editing on ten of the appendices. The end product is greatly improved because of his contribution.

Gratefully,

Kevin Forsberg, ESEP

Mike Krueger, CSEP

SEHv3 Contributions

The *INCOSE Systems Engineering Handbook* version 3 development team owes a debt of gratitude to all the contributors to prior editions (versions 1, 2, and 2a). The framework they provided gave a solid basis for moving ahead with this version. However this present document represents a significant departure from its predecessors since the goal was to create a shorter document consistent with the international standard ISO/IEC 15288:2002(E) – *Systems engineering–system life cycle processes*. As a result, we will not list all the contributors to earlier versions; interested readers are referred to the acknowledgment pages in those documents.

We want to thank the two co-chairs who worked with us in the formative stages of this handbook: John Leonard and Jim Chism. They provided valuable guidance and leadership in the difficult transition from a handbook based on the earlier versions to one based on ISO/IEC 15288.

It would be difficult to accurately characterize the specific contributions of each of the volunteers – section leads, steering committee, authors, and reviewers. Many served in multiple roles. A great deal of effort and enthusiasm was provided by the section leads and key authors, most of who also served on the steering committee. We acknowledge them in alphabetical order: Karen Bausman, Joe Carl, Sandy Friedenthal, Karl Geist, Ken Kepchar, Mike Krueger, Harold Kurstedt, Sean O'Neill, Mike Persson, Mary Redshaw, Andy Schuster, L. Mark Walker, and Jim Whalen. The steering committee also included the following people: Howard Eisner, Gerard Fisher, Richard Kitterman, David Long, William Mackey, and Paul Robitaille.

The review team led by Erik Aslaksen included in alphabetical order: Jonas Andersson, Lily Birmingham, Samantha Brown, John Clark, Michael Eagan, Ayman El-Fatatry, Patrick Hale, Jorg Lalk, Harold Lawson, Virginia Lentz, William Miller, Juan Miro, John Muehlbauer, Robert Peffer, Robert Porto, John Quitter, Tom Strandberg, Dan Surber, and David Walden. In addition, representatives from the INCOSE Hampton Roads Area Chapter, the Swedish Chapter, the Requirements Working Group, and the AIAA Technical Committee on Systems Engineering provided comments.

One of the requirements for this handbook is that it looks and reads as if it were written by a single person, and the reviewers all felt this objective has been met successfully. The co-chairs wish to thank our editor, Cecilia Haskins, for her dedication, and contributions to achieve this result.

We apologize in advance if we omitted anyone from this list in the final minutes before going to production.

Gratefully,

Terje Fossnes and Kevin Forsberg

Appendix F: Comment Form

Reviewed document: INCOSE SE Handbook v3.2.1

Name of submitter: Given FAMILY
(given name & family name)

Date Submitted: DD-MM-YYYY

Contact info: john.doe@anywhere.com
(email address)

Type of submission: group
(individual/group)

Group name and number of contributors: INCOSE XYZ WG
(if applicable)

Comments:

Detailed comments with reference to document section, paragraph, etc. Please include detailed recommendations, as shown in the table below.

Send comments to Info@incose.org

Comment ID	**Category** (TH, TL, E, G)	**Section Number** (e.g., 3.4.2.1, no alpha)	**Specific Reference** (e.g., Paragraph, Line, Figure, Table)	**Issue, Comment, and Rationale** (rationale must make comment clearly evident and supportable)	**Proposed Changed / New Text -- MANDATORY ENTRY --** (must be substantial to increase the odds of acceptance)

Note: TH = Technical High, TL = Technical Low, E = Editorial, G = General

This page intentionally left blank.

Index

Note: Many of the terms listed below appear with such frequency throughout the text that it was neither practical nor possible to list all occurrences. Therefore, this appendix identifies those occurrences corresponding to a definition or specific SE application of the term or concept of interest. In instances where an entire section or sequence of pages is devoted to a term or concept, only the first page of the series is identified.
